Leo Szilard: His Version of the Facts

The MIT Press
Cambridge, Massachusetts, and London, England

Leo Szilard: His Version of the Facts

Selected Recollections and Correspondence

Edited by Spencer R. Weart and
Gertrud Weiss Szilard

Volume II
All new material except that in the public domain
Copyright © 1978 by
Gertrud Weiss Szilard

All rights reserved.
No part of this book may be reproduced
in any form or by any means, electronic or mechanical,
including photocopying, recording, or by any information
storage and retrieval system, without permission
in writing from the copyright owner.

This book was prepared with
the support of NSF Grant SOC 75-00016. However,
any opinions, findings, conclusions, and/or recommendations herein
are those of the editors and do not necessarily reflect
the views of the NSF.

This book was set in Monophoto Times Roman,
printed and bound by
Halliday Lithograph Corporation
in the United States of America

First paperback printing, 1980

Library of Congress Cataloging in Publication Data

Szilard, Leo.
 Leo Szilard, his version of the facts.

 Bibliography: p.
 Includes index.
 1. Szilard, Leo—Interviews. 2. Szilard, Leo—
Correspondence. 3. Physicists—United States—
Correspondence. 4. Physicists—United States—
Interviews. 5. Physicists—United States—Biography.
I. Weart, Spencer R. II. Szilard, Gertrud Weiss.
QC16.S95A254 1978   539.7′092′4 [B]   78-15971
ISBN 0-262-19168-7 (hard)
ISBN 0-262-69070-5 (paper)

Ten Commandments
by Leo Szilard

1. *Recognize the connections of things and the laws of conduct of men, so that you may know what you are doing.*

2. *Let your acts be directed towards a worthy goal, but do not ask if they will reach it; they are to be models and examples, not means to an end.*

3. *Speak to all men as you do to yourself, with no concern for the effect you make, so that you do not shut them out from your world; lest in isolation the meaning of life slips out of sight and you lose the belief in the perfection of the creation.*

4. *Do not destroy what you cannot create.*

5. *(Untranslatable pun.)*

6. *Do not covet what you cannot have.*

7. *Do not lie without need.*

8. *Honor children. Listen reverently to their words and speak to them with infinite love.*

9. *Do your work for six years; but in the seventh, go into solitude or among strangers, so that the recollection of your friends does not hinder you from being what you have become.*

10. *Lead your life with a gentle hand and be ready to leave whenever you are called.*

Translated by Dr. Jacob Bronowski in June 1964

Zehn Gebote
von Leo Szilard

1. Erkenne die Zusammenhänge der Dinge und die Gesetze der Handlungen der Menschen, damit Du wissest, was Du tuest.

2. Deine Taten sollen gerichtet sein auf ein würdiges Ziel, doch sollst Du nicht fragen, ob sie es erreichen; sie seien Vorbild und Beispiel, nicht Mittel zum Zweck.

3. Sprich zu den Menschen wie zu Dir selber ohne Rücksicht auf die Wirkung Deiner Rede, auf dass Du die Menschen nicht ausstossest aus Deiner Welt und in der Vereinsamung der Sinn des Lebens Deinen Augen entschwindet und Du verlierest den Glauben an die Vollkommenheit der Schöpfung.

4. Zerstöre nicht, was Du nicht erschaffen kannst.

5. Rühre kein Gericht an, es sei denn, dass Du hungrig bist.

6. Begehre nicht, was Du nicht haben kannst.

7. Lüge nicht ohne Notwendigkeit.

8. Ehre die Kinder. Lausche andächtig ihren Worten und sprich zu ihnen mit unendlicher Liebe.

9. Verrichte sechs Jahre lang Deine Arbeit; im siebenten aber gehe in die Einsamkeit oder unter Fremde, damit die Erinnerung Deiner Freunde Dich nicht hindert zu sein, was Du geworden bist.

10. Führe das Leben mit leichter Hand und sei bereit fortzugehen, wann immer Du gerufen wirst.

30. Oktober 1940

Contents

List of Documents xi

Preface xvii

Acknowledgments xix

Note on the Text xxi

Chapter I
You don't have to be much cleverer, you just have to be one day earlier.

Recollections 3

Documents through January 1939 22

Chapter II
There was very little doubt in my mind that the world was headed for grief.

Recollections 53

Documents from December 1938 through July 1939 60

Chapter III
Sir: Some recent work by E. Fermi and L. Szilard... leads me to expect.... Yours very truly, A. Einstein.

Recollections 81

Documents from April 1939 through December 1939 87

Chapter IV
From that point on secrecy was on.

Recollections 115

Documents from February 1940 through October 1940   118

Chapter V
Somehow we did not seem to get the things done which needed to be done.

Recollections   143

Documents from December 1940 through February 1944   150

Chapter VI
Some of us began to think about the wisdom of testing bombs and using bombs.

Recollections   181

Documents from August 1944 through August 1945   189

Chapter VII
It was possible to tell people what we are facing in this century.

Recollections   223

Documents from August 1945 through December 1945   230

Source Notes   239

Name Index   241

List of Documents

1 Draft of Proposal for a New Organization Called "Der Bund" (about 1930) — 23
2 Letter to Sir William Beveridge (May 4, 1933) — 30
3 Letter to Dr. D. (May 7, 1933) — 32
4 Letter to Unknown Addressee (August 11, 1933) — 34
5 Letter to Lady Murray (April 24, 1934) with Attachment "Draft of Memorandum on the Sino-Japanese War" — 36
6 Letter to Sir Hugo [Hirst] (March 17, 1934) — 38
7 Memorandum of Possible Industrial Applications Arising out of a New Branch of Physics (July 28, 1934) — 39
8 Letter from M. Polanyi (November 11, 1934) — 40
9 Letter to F. A. Lindemann (June 3, 1935) — 41
10 Letter from M. Polanyi (June 28, 1935) — 43
11 Letter to M. Polanyi from Ch. Weizmann (July 5, 1935) — 43
12 Letter from Maurice Goldhaber (March 18, 1936) — 44
13 Letter to Niels Bohr (March 26, 1936) — 44
14 Letter to Lord Rutherford (May 27, 1936) — 45
15 Letter to J. Cockcroft (May 27, 1936) — 46
16 Cable to F. A. Lindemann (October, 1938) — 48
17 Letter to J. Tuck (October 21, 1938) — 48
18 Letter to F. A. Lindemann (January 13, 1939) — 50
19 Letter to Director of Navy Contracts (December 21, 1938) — 60
20 Cable to Director of Navy Contracts (January 26, 1939) — 60
21 Letter to Director of Navy Contracts (February 2, 1939) — 61
22 Letter to Lewis Strauss (January 25, 1939) — 62
23 Cable from F. A. Lindemann (February 3, 1939) — 63
24 Letter to Lewis Strauss (February 13, 1939) — 63
25 Letter from R. B. Kearney, Sales Manager, Radium Chemical Company (February 14, 1939) — 65
26 Letter from Edward Teller (about February 17, 1939) — 66
27 Letter to M. Tuve (March 22, 1939) — 67
28 Letter from M. Tuve (March 27, 1939) — 68

29	Letter to F. Joliot (February 2, 1939)	69
30	Cable to Lewis Strauss (February 19, 1939)	70
31	Cable to P. M. S. Blackett from V. Weisskopf (March 31, 1939)	70
32	Cable to H. von Halban from V. Weisskopf (March 31, 1939)	71
33	Letter to P. Dirac from E. Wigner (March 30, 1939)	71
34	Letter to P. M. S. Blackett from V. Weisskopf (about April 1, 1939)	72
35	Cable from F. Joliot, H. von Halban, L. Kowarski (April 5, 1939)	73
36	Cable to F. Joliot (April 6, 1939)	73
37	Letter to F. Joliot (April 7, 1939)	74
38	Cable from F. Joliot (April 6, 1939)	74
39	Cable to V. Weisskopf from P. M. S. Blackett (April 8, 1939)	74
40	Letter to Lewis Strauss (April 11, 1939)	74
41	Letter from M. Goldhaber (April 12, 1939)	75
42	Cable to P. M. S. Blackett (April 14, 1939)	76
43	Letter to P. M. S. Blackett (April 14, 1939)	77
44	Letter to Lewis Strauss (April 14, 1939)	77
45	Letter from Lewis Strauss (April 17, 1939)	78
46	Cable to F. Joliot (about April 14, 1939)	78
47	Letter from F. Joliot (April 19, 1939)	78
48	Letter to F. Joliot (July 5, 1939)	80
49	Letter from E. Wigner (April 17, 1939)	87
50	Letter to E. Wigner (May 21, 1939)	87
51	Letter to Lewis Strauss (July 3, 1939)	88
52	Letter from Ross Gunn (July 10, 1939)	89
53	Letter to A. Einstein (July 19, 1939)	90
54	Letter to A. Einstein (August 2, 1939)	92
55	Letter to President Roosevelt from A. Einstein (August 2, 1939)	94
56	Letter to A. Einstein (August 9, 1939)	96
57	Letter to E. Wigner (August 9, 1939)	97
58	Letter to A. Sachs (August 15, 1939)	97
59	Letter to Colonel Lindbergh (August 16, 1939)	99

60	Letter to A. Einstein (September 27, 1939)	100
61	Letter to A. Einstein (October 3, 1939)	101
62	Letter to Gano Dunn (September 13, 1939)	102
63	Letter from E. Wigner (September 26, 1939)	103
64	Letter to President Roosevelt from A. Sachs (October 11, 1939)	104
65	Letter to W. F. Barrett, Vice-President, Union Carbide and Carbon Corp., with Enclosure (October 18, 1939)	106
66	Letter to A. Einstein (October 17, 1939)	107
67	Letter to G. B. Pegram (October 21, 1939)	109
68	Letter to Lyman J. Briggs (October 26, 1939)	110
69	Letter to A. Sachs (November 5, 1939)	112
70	Letter to B. Liebowitz (December 4, 1939)	113
71	Letter to John T. Tate (February 14, 1940)	118
72	Letter to John T. Tate (April 5, 1940)	118
73	Letter to F. Joliot (April 12, 1940)	119
74	Letter to A. Einstein (March 7, 1940)	119
75	Letter to A. Sachs from A. Einstein (March 7, 1940)	120
76	Letter to President Roosevelt from A. Sachs (March 15, 1940)	121
77	Letter to A. Sachs from President Roosevelt (April 5, 1940)	122
78	Letter to A. Sachs (April 22, 1940) with Memorandum	123
79	Letter to Lyman J. Briggs from A. Einstein (April 25, 1940)	125
80	Letter from Louis A. Turner (May 27, 1940)	126
81	Letter to Louis A. Turner (May 30, 1940)	127
82	Memorandum for H. Urey (May 30, 1940)	129
83	Memorandum for A. Sachs (undated)	131
84	Letter from Louis A. Turner (June 1, 1940)	132
85	Letter from G. Breit (June 5, 1940)	133
86	Letter to E. Fermi (July 4, 1940)	133
87	Letter to G. Breit (July 6, 1940)	135
88	Letter to E. Wigner (July 6, 1940)	136
89	Letter to A. Sachs (August 28, 1940) with Draft of Memorandum (August 27, 1940)	137

90	Letter to G. B. Pegram (about October 1940)	139
91	Letter to V. C. Hamister, National Carbon Company (December 16, 1940)	150
92	Letter to H. D. Batchelor, National Carbon Company (February 7, 1941)	150
93	Letter to V. Bush (May 26, 1942)	151
94	Letter from V. Bush (June 1, 1942)	153
95	"What is Wrong with Us?" (September 21, 1942)	153
96	Memorandum to A. H. Compton on "Compartmentalization of Information and the Effect of Impurities of 49" (November 25, 1942)	160
97	Letter to V. Bush (January 14, 1944)	161
98	"Proposed Conversation with Bush," Parts I–IV (February 28, 1944)	164
99	Memorandum (August 10, 1944)	189
100	Letter to Lord Cherwell (August 18, 1944)	192
101	"Atomic Bombs and the Postwar Position of the United States in the World" (Spring, 1945)	196
102	Letter to President Roosevelt from A. Einstein with Enclosure by Szilard (March 25, 1945)	205
103	Letter to President Truman (May 25, 1945)	208
104	Letter from E. Teller (July 2, 1945)	208
105	Letter to Group Leaders of the Metallurgical Laboratory (July 4, 1945)	209
106	Reply by Group Leaders of the Metallurgical Laboratory (July 13, 1945)	210
107	A Petition to the President of the United States (July 17, 1945)	211
108	Letter to E. Creutz (July 10, 1945)	212
109	Letter to E. Wigner (July 7, 1945)	213
110	Letter to A. H. Compton (July 19, 1945)	214
111	Memorandum to Colonel K. D. Nichols from A. H. Compton (July 24, 1945)	214
112	Letter to Matthew J. Connelly (August 17, 1945)	215
113	Cable from Matthew J. Connelly (August 25, 1945)	216
114	Letter from James S. Murray, Captain, Corps of Engineers (August 27, 1945)	216
115	Letter from James S. Murray (August 28, 1945)	219
116	Letter to R. M. Hutchins (August 29, 1945)	220
117	Letter to Alfred W. Painter (August 11, 1945)	230

List of Documents

118	Proposed Petition to the President of the United States (August 13, 1945)	231
119	Draft of a Platform for Conversations with Congress (September 7, 1945)	231
120	Address to the Atomic Energy Control Conference, University of Chicago (September 21, 1945)	233
121	Letter to William Benton (October 5, 1945)	235
122	Cable from William Benton to James B. Conant (December 22, 1945)	237

Preface

Leo Szilard was a scientist who, not content to remain in the laboratory, went beyond its walls to attend to the larger causes of humanity. External crises several times interrupted his scientific career, as would be expected for anyone of Central European ancestry born in Budapest in 1898. But even after he came to the relative security of the United States in 1938, his career was often interrupted, by his own will: he pressed the Americans to make nuclear weapons before Nazi Germany did; he was then among the first to fight against the actual use of the bombs he had done much to create; he switched from nuclear physics to biology in order to keep abreast of the moving edge where scientific discovery vitally affects mankind; until his sudden death in California in 1964 he threw himself time and again into schemes to advance world peace. Beneath all this turbulent movement he held confidently and with absolute consistency to his single goal—creating what he called "a more livable world."

At various times Szilard considered writing his autobiography, but his interest was always captured by the present and the future. Therefore he was not inclined to spend much time on the past and never wrote more than fragments. He had a sense of history, however, and carefully preserved correspondence and other documents which he considered historically significant.

In 1951 Szilard contemplated writing a history of the Manhattan Project and organized pertinent documents and drafted some notes, among which we found the following anecdote. While talking to a colleague about some disturbing things that had happened during the project, Szilard said that he was going to write down the facts, not for publication, just for the information of God. When his colleague remarked that God might know the facts, Szilard replied that this might be so, but "not *this* version of the facts."

This volume then presents "Szilard's Version of the Facts" through tape-recorded interviews, supplemented by correspondence and other documents. Most of the tapes were recorded in 1960 during a period of serious illness which kept him in the hospital for a year. At that time a tape recorder was put into his room supposedly for the purpose of recording his memoirs. Being interested more in the future than in the past, he used it chiefly to dictate the first draft of the utopian story "The Voice of the Dolphins."[1] However, on a few occasions he responded with zest to questions about his past, and then the tape recorder was switched on. Other reminiscences were recorded in 1956 and 1963. The tapes found in the Szilard files, many transcribed posthumously, deal with the period from his childhood to the year 1946 only. We have tried to use only a minimum of scholarly apparatus and let Szilard himself speak. A previous volume, containing his scientific papers, was published several years ago.[2]

This volume covers the years from Szilard's childhood through his early reaction to

---

[1]Leo Szilard, *The Voice of the Dolphins and Other Stories* (New York: Simon and Schuster, 1961).
[2]Bernard T. Feld and Gertrud Weiss Szilard, eds., *The Collected Works of Leo Szilard. Scientific Papers* (Cambridge, Mass., and London, England: The MIT Press, 1972).

the dropping of the atomic bombs. Until that point he had worked largely behind the scenes, but afterward he devoted much of his attention to public speaking and writing. To keep this volume to a manageable size, we have brought it to an end at the period 1945–1946 when he became a public figure. Abundant social, political, and literary writings that originated after Hiroshima remain to be published.

In presenting his writings, we hope that Szilard's voice will continue to be heard because we believe that he still has much to tell us. The recollections are fragmentary and should not be taken as a complete autobiography, but they do reveal the essence of Szilard's thinking and approach to life. Szilard was less interested in his own personal situation than in urgent social and political problems and in the problems of others. In editing the writings, we have tried to follow his pattern.

In 1940, in a semiserious mood, Szilard formulated his own "Ten Commandents." He wrote them in German and was never satisfied with any attempts at translation. They were therefore published only in the German edition of his book *The Voice of the Dolphins*.[3] (They were translated posthumously for distribution to his friends.) We have inserted them here as a frontispiece because they reflect Szilard's spirit like a portrait.

Spencer R. Weart
American Institute of Physics
Center for History of Physics

Gertrud Weiss Szilard
Program in Science, Technology,
and Public Affairs
University of California, San Diego

[3]Leo Szilard, *Die Stimme der Delphine* (München: Wilhelm Heyne Verlag, 1976).

Acknowledgments

This work was supported by grants from the National Science Foundation[1] and the Cyrus Eaton Foundation, which we gratefully acknowledge. We also wish to thank the American Institute of Physics Center for History of Physics for their assistance.

The project was sponsored by the Program in Science, Technology, and Public Affairs of the University of California, San Diego. Special thanks are due to Herbert York, the director of the program, who not only provided space and other supportive services but was always available for consultation and gave advice and assistance whenever needed. Without his help the project could not have been carried out.

We also wish to thank Melvin Voigt, university librarian recently retired, and his successor Millicent Abell for providing safe space for the Szilard archives and use of the library services as well as for their expert advice on technical matters. Doris Hamstra, Office of Graduate Studies and Research, gave valuable assistance in administrative and personnel matters to the project.

Helen Hawkins prepared the draft manuscript of Szilard's collected post-Hiroshima papers, yet to be published, and assisted in preparing the documents and annotations of the final chapter of this volume. We also wish to acknowledge the assistance given by Kathleen Winsor in preparing a preliminary version of the autobiographical fragments and a compilation of documents.

For permissions to include unpublished letters we thank Gregory Breit, Maurice Goldhaber, C. R. Gunn, Warren M. Holm, Pierre Joliot, John C. Polanyi, Lewis H. Strauss, Edward Teller, Louis A. Turner, Merle Tuve, Meyer W. Weisgal, and Eugene P. Wigner.

[1]NSF Program in History and Philosophy of Science.

Note on the Text

The text of the recollections is drawn largely from transcripts of the taped interviews interspersed with material from other sources, arranged in a chronological sequence. The transcripts required only the minimum editing necessary to change spoken to written grammar. Some of the material taped in 1960 was published as "Reminiscences" in *Perspectives in American History, 2* (1968), pp. 94–151.[1] The present version omits none of this, but is edited somewhat closer to the original language to give the flavor of Szilard's speech at the expense of some grammatical finish. A few paragraphs have been moved to put them in chronological order, and further material from other sources has been incorporated. The source of each segment in the text is indicated at the end of the segment by an italic number in brackets. The key to the sources is given on p. 239.

In addition to these recollections Szilard's unpublished social and political writings for this period are documented in many letters and memoranda. We have divided the recollections into chapters and provided a selection of documents to complement the text. We have chosen particularly those documents Szilard himself selected as important and filed in folders marked "History"; other documents were selected to show the range of his social and political thought; and a few less important but typical examples are also included. The documents are numbered consecutively and are generally in chronological order, except where related items are together.

We have worked chiefly from materials in Szilard's files, which are complete enough to give a representative picture, at least from the early 1930s on. Since we have often relied on carbon copies, we are not certain that all the letters were sent in precisely the form given, although this does seem nearly always to have been the case. We have also included some drafts that were not sent, for our chief aim is to illustrate Szilard's thinking.

Documents originally written in a foreign language (French, German, Hungarian) are given in English translation. In a few instances where a document was particularly significant, as in the Szilard-Einstein correspondence (in German) of 1939, we also show the original. Document titles are Szilard's unless otherwise indicated by brackets. In the list of documents we have listed some documents by more descriptive titles.

The editors' work on the text is confined to six categories: (1) Headnotes, always in *italics*, to put Szilard's words in context. In some instances one headnote refers to several following documents. (2) Footnotes, always the editors' rather than Szilard's, to explicate obscure points and to identify references and people. People are identified by their position at the time Szilard mentions them; the reader should use the index, since names are sometimes identified only by context and not necessarily at their first appearance. Citations in footnotes to editorial material are to documents in the Szilard files. (3) Silent corrections of obvious and trivial errors in spelling, punctua-

---

[1]Reprinted in Donald Fleming and Bernard Bailyn, eds., *The Intellectual Migration: Europe and America, 1930–1960* (Cambridge, Mass.: Harvard University Press, 1969).

tion, and grammar; in particular we have often put those parts of Szilard's oral recollections narrated in the present tense into a past tense. (4) Substantive additions and grammatical replacements in the text, always in square brackets, and deletions, always marked by an ellipsis. Signatures are added to letters Szilard wrote, even when absent from our copy. (5) When other sources make it possible, dates for documents undated in the original are supplied. For the reader's convenience, when a letter is written by someone other than Szilard and this is not immediately obvious on first sight, the writer's name is supplied in square brackets. (6) References to documents printed in this volume are indicated by italics, thus: *Doc. 1*. The list of documents shows where the documents may be found.

Addresses and other incidentals of letters where given on the copy in Szilard's files are printed. Telegrams are printed entirely in capitals.

Leo Szilard: His Version of the Facts

Chapter I
You don't have to be much cleverer, you just have to be one day earlier.

As far as I can see, I was born a scientist. I believe that many children are born with an inquisitive mind, the mind of a scientist, and I assume that I became a scientist because in some ways I remained a child.

Very often it is difficult to know where one's set of values comes from, but I have no difficulty in tracing mine to the children's tales which my mother used to tell me. My addiction to the truth is traceable to these tales and so is my predilection for "Saving the World." [1]* My mother was fond of telling tales to her children and she always had some particular purpose in mind. Why she wanted to inculcate addiction to truth in her children is not clear to me.

I remember one story, which made a deep impression on me, about my grandfather. My grandfather was a high school student at the time of the Hungarian Revolution in 1848. In high school, when the children were waiting for the teacher to turn up, it was customary in Hungary for one child to keep watch. It was his task to keep a list of those children who were disorderly, and when the teacher came to class he was supposed to submit the list of these disorderly children to the teacher for punishment. In the particular case of the story my mother told me, the Hungarian Revolution of 1848 was on. A troop of soldiers was marching by the school and a number of children violated orders by leaving the class and lining the street and cheering the soldiers. My grandfather, who was supposed to keep watch on disorderly children, joined those who left the school building and cheered the soldiers. When the teacher turned up for class, all the children were back in the classroom and my grandfather rendered his report. He gave the teacher the list of those children who violated orders and went out to the street, and this list included his own name. The teacher was so much taken aback by this frankness that nobody was punished. [2]

Apart from my mother's tales the most serious influence on my life came from a book which I read when I was ten years old. It was a Hungarian classic, taught in the schools, *The Tragedy of Man*.[1] I read it much too prematurely and it had a great influence on me, perhaps just because I read it prematurely. Because I read it, I grasped early in life that "it is not necessary to succeed in order to persevere."[1]

I remember that I was a very sensitive child and somewhat high-strung. I couldn't say that I had a happy childhood, but my childhood was not unhappy either. For some

---
*See Source Notes, p. 239.
[1] By Imre Madach; a metaphorical dramatic poem finished in 1862. In an interview reported in the *New York Post*, November 24, 1945, Szilard described the novel as follows: "In that book the devil shows Adam the history of mankind, with the sun dying down. Only Eskimos are left and they worry chiefly because there are too many Eskimos and too few seals. "The thought is that there remains a rather narrow margin of hope after you have made your prophecy and it is pessimistic."

reason or other I was frequently ill up to the age of ten, and I did not go much to school. I had mostly instruction at home. [2]

I was the oldest of three children, and we lived in a house which belonged . . . originally to my grandparents. Then it was inherited by three sisters, of whom my mother was one, and each sister had a whole floor. It was a house with a large garden in the cottage district of Budapest. . . . I remember that I was already very intensely interested in physics when I was thirteen. At that time I got a few playthings in physics, and I remember how overjoyed I was. [3]

From my tenth year I was sent to a public school. For some reason or other, throughout the eight years which I spent at the public school (until I reached the age of eighteen), I was always a favorite of the class. Just precisely what this was due to, I couldn't say. I suppose it was somehow the reaction of the class to my personality, and I somehow cut a favorite figure from the point of view of a set of values which were at that time prevalent in the city of Budapest. There were others who had better marks in school, even though I had pretty good marks, but these others obviously strove to get good marks, and this was resented by the class. My good marks simply came from the fact that I had no difficulty in keeping ahead. I was interested in science, I was interested in mathematics, and I knew languages because we had governesses at home, first in order to learn German and second in order to learn French.

Perhaps my popularity was also due to my frankness, which was coupled with a lack of aggression. One of the favorite sports of the class at that time was playing soccer. I was not a good soccer player, but because I was liked there was always a rivalry between the two teams: On whose side would I be? I was sort of a mascot. They discovered early that I was, from an objective point of view, no asset to the team, and it didn't take them long to discover that I could do least damage by being the goalkeeper. So up to the age of fifteen, when I finally refused, I played every soccer game of the class on one or the other side, very often on the losing side. In thinking back, I have a feeling of gratitude for the affection which went so far that my classmates did not mind occasionally losing a game for the sake of having me on their team.

I must have made a rather strong impression on my schoolmates, judging from the fact that they reported to me years later conversations which they had had with me and which I had forgotten. One of these "memorable" conversations occurred at the outset of the First World War. I was sixteen at the time, and when the war started we didn't have a very good conception of what kind of an enterprise this was. Most people thought that the war would last just a few months and, as the German Kaiser once said, our troops would be back by Christmas. He meant Christmas, 1914.

There was speculation in the class as to who might win the war, and apparently I said to them at the time that I did of course not know who would win the war, but I did know how the war ought to end. It ought to end by the defeat of the central powers, that is the Austro-Hungarian monarchy and Germany, and also end by the defeat of Russia. I said I couldn't quite see how this could happen, since they were fighting on opposite sides, but I said that this was really what ought to happen. In retrospect I find it difficult to understand how at the age of sixteen, and without any direct knowledge of countries other than Hungary, I was able to make this statement. Somehow I felt that Germany and the Austro-Hungarian Empire were weaker political structures than both France and England. At the same time I felt that Russia was a weaker political structure than was the German Empire.

I am inclined to think that my clarity of judgment reached its peak when I was sixteen, and that thereafter it did not increase any further and perhaps even declined. Of course, a man's clarity of judgment is never very good when he is involved, and as you grow older, and as you grow more involved, your clarity of judgment suffers. This is not a matter of intelligence; this is a matter of ability to keep free from emotional involvement.

I was certainly remarkably free of emotional involvement during the First World War. I remember that when the war started we were at an Austrian resort in Velden. We immediately made arrangements for returning to Budapest by train, and as our train moved slowly towards Budapest, more and more troop trains pulled alongside the train or passed us. Most times, the soldiers in all these trains were drunk. Some of the fellow passengers, looking out of a window and seeing the troop trains pass by, made a remark to my parents that it was heartening to see all this enthusiasm; and I remember my comment, which was that I could not see much enthusiasm but I could see much drunkenness. I was immediately advised by my parents that this was a tactless remark, which I am afraid had only the effect that I made up my mind then and there that if I had to choose between being tactless and being untruthful, I would prefer to be tactless. Thus my addiction to the truth was victorious over whatever inclination I might have had to be tactful. [2]

The set of values of the society in which I lived in Budapest was conducive for a young man to dedicate himself to the pursuit of science, and the poor quality of the teaching of science at the universities in Hungary furnished stimulation to independence of thought and originality. [1] [In 1916,] one year before I was drafted, I entered the Hungarian institute of technology as a student in order to study electrical engineering.[2] My real interest at that time was physics, but there was no career in physics in Hungary. If you studied physics, all that you could become was a high school teacher of physics—not a career that had any attraction for me. Therefore I considered seriously doing the next best thing and studying chemistry. I thought that if I studied chemistry I would learn something that was useful in physics and I would have enough time to pick up whatever physics I needed as I went along. This I believe in retrospect was a wise choice. But I didn't follow it, for all those whom I consulted impressed upon me the difficulty of making a living even in chemistry, and they urged me to study engineering. I succumbed to that advice, and I cannot say that I regret it, because whatever I learned while I was studying engineering stood me in good stead later after the discovery of the fission of uranium.

The war years were rather uneventful for me, even though one year before the end

---

[2]Kir. Jozsef-müegyetem in Budapest, the King Joseph Institute of Technology.

of the war I was drafted into the army. In Austria and Hungary, again corresponding to the set of values of those times in those localities, a young man who had high school education was automatically scheduled to become an officer, so I was sent to officers' school. And again in accordance with the set of values of those times, I ended up third in the officers' school of the brigade in spite of my rather unmilitary posture. Even though I was obviously not what you might call a good soldier, my teachers were impressed with my ability to grasp scientific and technical problems. Because I was able to explain how the telephone worked when nobody else in the class could explain the functioning of this mysterious gadget, I had a certain amount of standing in the class; and in spite of my unmilitary behavior I ended up third in my class, which comprised the officers-to-be of that particular year.

Since people have no imagination whatsoever, they cannot imagine in peacetime that there should be war, and if the war goes on for a few years, they cannot imagine that there ever will be peace. So, by the time of the third year of the war when I was drafted and sent to officers' school, nobody could imagine that war would ever end, and therefore more and more emphasis was put on the thorough training of officers. So after I left officers' school, we were sent for further training to a camp which was established on the German frontier in Kufstein. There, while war was raging elsewhere, we were taken on daily trips to the mountains, the Kaisergebirge, and were trained in other completely superfluous activities. I imagine that we were in reserve, and that at that particular moment there was no great need at the front for officers in the Austro-Hungarian Army.

Our commanding officer, a Captain, had only a few fixed ideas which bothered us; otherwise, he left us complete freedom. One of these ideas was that it was not becoming to an officer to walk along the street with his gloves in his hands. He should either wear no gloves or he should put the gloves on. He was also concerned about our being properly dressed, and he was indignant when he learned that we didn't bring with us our dress uniforms. From then on it became customary to ask for leave of absence to go home in order to pick up one's dress uniform. If the war had lasted long enough we would all have ended up with our dress uniforms, ready for festive occasions. Those who went home to Hungary to pick up their dress uniforms were also expected to bring some flour. There was a shortage of foodstuffs in Austria while Hungary still had plentiful food. These leaves of absence usually amounted to about one-fifth to one-third of the school being absent on trips home.

One day I awoke with a strong headache and high temperature. This frightened me because I knew that if I came down with something like pneumonia I would be put in a hospital in Kufstein, and I would never have a chance to be sent to a hospital back home. Rules in this respect were very severe. We were in the so-called Etappe [Communications Zone] and nobody who was ill could go home. I therefore decided that I would ask for a leave of absence to go home for a few days, and then report ill if my condition got worse. This way I would land in a hospital in Budapest near home, and if I were seriously ill I could have my family look after me. In order to get a leave of absence it was necessary to go through the routine of reporting to the commanding officer in a ceremonial form. Those who had any requests or who were ordered to face punishment, usually fifteen to twenty on any one day, were lined up in a corridor of the office building where the commanding officer resided and had to await his pleasure. After a half an hour or an hour he would turn up, stop in front of each man, and each man could bring forward his request. My main difficulty was that by the time this formality started, around noon, my temperature was 102°F. Standing at attention for half an hour or an hour, or even standing relaxed, is rather a strain if you have a high temperature, and there is always a danger that you may fall on your face. However, I somehow pulled myself together, and when my turn came to speak up to the commanding officer I asked him for a leave of absence to go home for a week, because my brother had a serious operation and my parents needed my moral support. He said that he had no objection to my getting a leave of absence and to my going home for a week, but right now there were just too many people on leave of absence, and I ought to wait until a few people had come back, and then he would give me leave to go. I immediately replied that the operation of my brother could not be postponed, and that if it were impossible for me to get a leave of absence now for a week, then I would modify my request and ask for a leave for two days so that I could be home on the day of the operation. The Captain was taken aback, because while it was perfectly all right to lie, it was not customary to insist if the request was refused. However, just because he was taken aback and didn't know quite what to think, I got my leave of absence.

Now the difficulty was how to get to the train, which left about midnight. With the support of a few of my comrades who kept me erect, I was finally pushed into the train and sat down in a corner in a compartment. There were a few other officers in the same compartment, and when morning came one of them told me, "Do you feel better now? You were pretty drunk last night." "I was not drunk," I told him, "I was ill." He didn't reply to this but I could see that he didn't believe me.

As the train approached Vienna I took my temperature again, and to my horror I saw that it was falling. I spent the night in Vienna and asked a doctor to look me over. He told me that I had no pneumonia and I was not in bad shape. The next day I arrived in Budapest with my temperature down but with a persistent cough making its appearance. That I landed in a hospital in Budapest was not due to my state of health but to my family connections. I wrote to my commanding officer, expressing my regret that I was not able to return to the cause, and got an affectionate letter commending me for my past military performance and wishing me good luck.[3] A week later I received another letter advising me that the class had been dissolved and that everybody had been sent to the front.

Two weeks later I had a letter from my commanding officer, advising me that an

---

[3]Letter datelined K.u.k. Geb. Artill. Säbelchargenkurs, Kufstein, October 10, 1918, in Szilard Papers.

epidemic of Spanish flu had broken out in the school and that the school was practically closed. It appears that I may have been the first victim of the Spanish influenza in the school and perhaps in the whole Austrian Army.

Not long afterwards, I heard that my own regiment at the front had been under severe attack and that all of my comrades had disappeared. So it appears that the Spanish flu (which caused the death of many of my comrades) saved my own life. (Perhaps I should add to the Spanish flu my own determination to go home when I was ill.)

The collapse of the Austro-Hungarian army was followed by a troubled period in Hungary that ended with the Communist government of Bela Kun, which lasted about four months [summer, 1919, at which time Szilard had returned to the Institute of Technology]. This government lasted too short a time to be able to do anything except hold office. During this period the things which had deteriorated during the war deteriorated even further, and I made up my mind that I wanted to leave Hungary to study in Germany.

During the troubled times of the Communist regime of Bela Kun I made a strenuous effort to obtain a passport to continue my studies of electrical engineering in Germany. One or two days after these efforts were successful the regime collapsed and was replaced by the regime of Horthy.[4] Thus I had to start from scratch in my quest for a passport, but through the help of friends I got one rather quickly, and I left Hungary [around Christmas, 1919] to go by way of Vienna to Berlin. This was about the worst time after the war because of the coal shortage. There was a shortage of food and there was a shortage of coal; because of the shortage of coal travel was slow, and as a matter of fact it took me two weeks to get from Budapest through Vienna to Berlin.

I stayed in Vienna for only a few days—as long as was necessary to make arrangements for the trip to Berlin. But during those few days I was greatly impressed by the attitude of the Viennese, who in spite of starvation and misery were able to maintain their poise and were as courteous as they have always been, to each other as well as to strangers.

In Berlin I had to face new difficulties. The number of foreign students who were admitted was limited. The attitude towards foreign students was not friendly in this respect. . . .[5] I applied for admission to the Technische Hochschule of Berlin. This permission I finally got, but not without difficulty and not without having to bring to bear all the pressure I could through such private connections as I was able to muster in the city of Berlin.

Berlin at that time lived in the heyday of physics. [Albert] Einstein was there, Max Planck and [Max] von Laue were at the University of Berlin, and so was Walter Nernst; and Fritz Haber was at that time director of one of the Kaiser Wilhelm Institutes. Engineering attracted me less and less, and physics attracted me more and more,

---

[4]Bela Kun's regime collapsed August 1, 1919; there followed the occupation of Budapest by the Rumanian army, and on March 1, 1920, the rule of Nicolas Horthy as regent.
[5]We omit a paragraph describing how, after difficult maneuvers, Szilard obtained a German visa.

and finally the attraction became so big that I was physically unable to listen to any of the lectures through which I sat, more or less impatiently, at the Institute of Technology.

Even though all arguments mustered by the conscious spoke in favor of getting a degree in engineering rather than getting a degree in physics, whatever considerations went on at the subconscious level argued for the opposite. In the end, as always, the subconscious proved stronger than the conscious and made it impossible for me to make any progress in my studies of engineering. Finally the ego gave in, and I left the Technische Hochschule to complete my studies at the University, some time around the middle of '21. (Figure 1)

A student of physics had great freedom in those days in Berlin. Boys left high school when they were eighteen years old. They were admitted at the University without any examinations. There were no examinations to pass for four years, during which time the student could study whatever he was interested in. When he was ready to write a thesis, he either thought of a problem of his own or he asked his professor to propose a problem on which he could work. At the better universities, and Berlin belonged to them, a thesis in order to be acceptable had to be a piece of really original work. If the thesis showed the student to be really able and was accepted, the student had to pass an oral exam. [2]

At some point, rather early, I went to von Laue, who was professor of theoretical physics, and asked him whether he would give me a problem on which I could work to get my doctor's degree. . . . I had this problem [in the theory of relativity] which von Laue gave me, but I couldn't make any headway with it. As a matter of fact, I was not even convinced that this was a problem that could be solved. I forced myself to work on it, but it just wouldn't go at all. This went on for about six months. Then came Christmas 1921; and I thought Christmastime is not a time to work, it is a time to loaf, so I thought I would just think whatever comes to my mind. Pretty soon things began to come into my mind in a field completely unrelated to the theory of relativity.

I went for long walks and I saw something in the middle of the walk; when I came home I wrote it down; next morning I woke up with a new idea and I went for another walk; this crystallized in my mind and in the evening I wrote it down. There was an onrush of ideas, all more or less connected, which just kept on going until I had the whole theory fully developed. It was a very creative period, in a sense the most creative period in my life, where there was a sustained production of ideas.[6] Within three weeks I had produced a manuscript of something which was really quite original. But I didn't dare take it to von Laue, because it was not what he asked me to do.

There was a seminar for students which Einstein held at the time, which I attended, and after one of these seminars I went to him and said that I would like to tell him about something I had been doing. He said, "Well, what have you been doing?" And I told him what I had done. And Einstein said, "That's impossible. This is something

---

[6]In the original this story, beginning with "I went for long walks . . ." was told a little later, after ". . . a cornerstone of modern information theory."

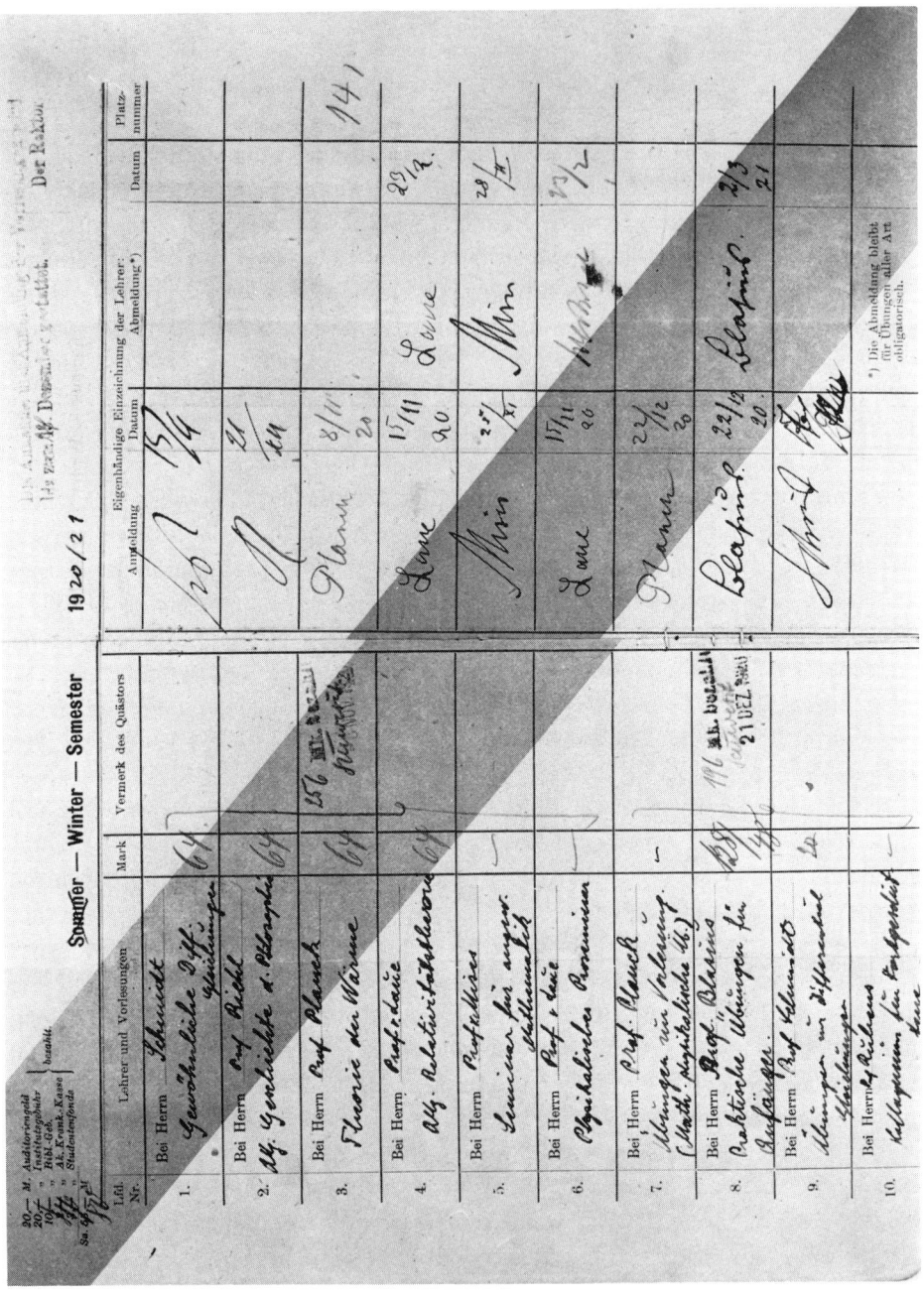

Figure 1
A record of courses Szilard attended at the University of Berlin in the winter semester 1920/21. The courses are listed in Szilard's handwriting and, as was usual in German universities, the professors signed the booklet personally in the beginning and at the end of the semester certifying that the student had actually attended.

that cannot be done." And I said, "Well, yes, but I did it." So he said, "How did you do it?" It didn't take him five or ten minutes to see, and he liked this very much. This then gave me courage and I took the manuscript to von Laue. I caught him as he was about to leave his class and I told him that while I had not written the paper which he wanted me to write, I had written something else, and I wondered whether he might be willing to read it and tell me whether this could be used perhaps as my dissertation for the Doctor's degree. He looked somewhat quizzically at me, but he took the manuscript. And next morning, early in the morning, the telephone rang. It was von Laue. He said, "Your manuscript has been accepted as your thesis for the Ph.D. degree."[7]

Up to the time that I wrote this thesis it was generally believed that the laws which govern the thermodynamical fluctuations must be derived from mechanics, and that they transcend what is called the second law of thermodynamics. I showed that the second law of thermodynamics was much more than just a plain statement about the average values; it also covers the laws which govern the thermodynamic fluctuations. This was not really a beginning, it was not the cornerstone of a new theory, it was rather the roof of an old theory.

However, about six months later I wrote a little paper which was on a rather closely related subject.[8] It dealt with the problem of what is essential in the operations of the so-called Maxwell's Demon, who guesses right and then does something, and by guessing right and doing something he can violate the second law of thermodynamics. This paper was a radical departure in thinking, because I said that the essential thing here is that the demon utilizes information—to be precise, information which is not really in his possession because he guesses it. I said that there is a relationship between information and entropy, and I computed what this relationship was. No one paid any attention to this paper until, after the war, information theory became fashionable. Then the paper was rediscovered. Now this old paper, to which for over 35 years nobody paid any attention, is a cornerstone of modern information theory....

In maybe 1928 or 1929 I began to think what might be the future development in physics. Disintegration of the atom required higher energies than were available up to that time; there had been no artificial disintegration of the atom. I was thinking of how could one accelerate particles to high speeds. I hit upon the idea of the cyclotron, maybe a few years before Lawrence.[9] I wrote it down in the form of a patent application which was filed in the German patent office.[10] It was not only the general idea of the cyclotron, but even the details of the stability of the electron orbits, and what it would take to keep these orbits stable; all this was worked out on this occasion. [3]

---

[7]Szilard, *Zeitschrift für Physik*, *32*: 753–788 (1925), in Bernard T. Feld and Gertrud Weiss Szilard, eds., *The Collected Works of Leo Szilard. Scientific Papers* (Cambridge, Mass.: MIT Press, 1972), pp. 34–69, hereinafter referred to as *Scientific Papers*.
[8]Szilard, *Zeitschrift für Physik*, *53*: 840–856 (1929), in *Scientific Papers*, pp. 103–119. For a discussion see Carl Eckart's introduction, ibid., pp. 31–33.
[9]Ernest O. Lawrence, physicist at the University of California, Berkeley, first to build a working cyclotron (1932).
[10]Application S.89288, filed January 5, 1929, in *Scientific Papers*, pp. 554–563.

In the meantime jointly with Professor Albert Einstein I thought of a method of pumping liquid metals through tubes with the action of a moving magnetic field on electric currents induced by this field in the liquid metal. The German General Electric Company (A.E.G.) wanted to develop a pump based on this principle and for about three years I acted as a consultant to them for this development. [4] This was patented also, again in Germany.[11] We wanted to use it to make a household refrigerator without moving parts and, as a matter of fact, we built one refrigerator which was based on this principle. It was not very practical, because mechanical refrigerators which have moving parts function really quite well and are not too noisy [whereas the liquid metal pump was very noisy]. [3] In 1932 the A.E.G. discontinued this development and not until the introduction of atomic reactors in 1942 did this system find application. [4]

Einstein is a great man and this manifests itself in any number of ways. He is a man as free from vanity as I have ever seen. I heard him talk to an audience of a thousand in German where he is at his best, and he talked to them as he would talk to a few friends gathered at the fireside. That does not mean that he is not shy, of course. Great achievement in the field of science requires a certain kind of sensitivity, and sensitivity leads to shyness.

If you meet him you are struck with his great modesty and his great simplicity of heart. This simplicity is perhaps the key to the understanding of his work—in science the greatest thoughts are the simplest thoughts. He started out as an examiner in the Swiss Patent Office. There he earned a living not dependent on giving birth to any scientific ideas great or small. "Why don't you take a job in the Patent Office?" Einstein told me when I had just graduated. "That would be best for you; it is not a good thing for a scientist to be dependent on laying golden eggs. When I worked in the Patent Office, that was my best time of all."

Einstein has always taken an interest in the affairs of the world, a remote sort of interest, ready to withdraw at an instant's notice into the nonsecular world of science. "One-fifth of my heart I am ready to spend on the world," he said to me once, "but four-fifths of it I retain. I could of course spend all my time buttonholing people and talk to them about the inevitable downfall of the capitalist system—if you please—or make speeches on the street corner. Maybe they would put me in jail, but what good would I do? No," he said, "Not everyone is as fortunate as Christ. To sacrifice yourself and do some good, that takes luck." If you are willing to use the words "religious attitude" in the broad sense of the term, then Einstein is a religious man in the deepest sense of the word. "As long as you pray to God and ask him for *something*, you are not a religious man," he said to me—and I share his view. [5]

In 1932 my interest shifted to nuclear physics and I moved to the Harnack House in

---

[11]Einstein and Szilard shared seven German patents on pumps and refrigeration systems, taken from 1927 through 1930. See *Scientific Papers*, pp, 540–541, 701–705. Szilard also held a variety of other patents, taken from 1923 on.

Berlin-Dahlem with the thought of taking up some experimental work in one of the Kaiser Wilhelm Institutes there. I discussed the possibility of doing experiments in nuclear physics with Miss Lise Meitner[12] in the Kaiser-Wilhelm-Institut für Chemie, but before this led to a final discussion one way or the other, the political situation in Germany became tense and it seemed advisable to delay a final decision. [4]

I reached the conclusion something would go wrong in Germany very early. (*Doc. 1*) I reached this conclusion in 1930, and the occasion was a meeting in Paris. It was a meeting of economists who were called together to decide whether Germany could pay [war] reparations, and just how much reparations Germany could pay. One of the participants of that meeting was Dr. Hjalmar Schacht, who was at that time, I think, president of the German Reichsbank. To the surprise of the world, including myself, he took the position that Germany could not pay any reparations unless she got back her former colonies.[13] This was such a striking statement to make that it caught my attention, and I concluded that if Hjalmar Schacht believed he could get away with this, things must be rather bad. I was so impressed by this that I wrote a letter to my bank and transferred every single penny I had out of Germany into Switzerland. I was not the only one, as I later learned. Within a few months after this speech of Schacht's, a very large sum of money was drawn out of Germany, mainly by depositors from abroad. Apparently there are many people who are sensitive to this kind of signal.

I visited America in 1931. I came here on Christmas Day, 1931, on the *Leviathan*, and stayed here for about three months.[14] In the course of 1932 I returned to Berlin where I was *Privatdozent* at the University. Hitler came into office in January '33, and I had no doubt what would happen. I lived in the faculty club of the Kaiser Wilhelm Institute in Berlin-Dahlem and I had my suitcases packed. By this I mean that I had literally two suitcases standing in my room which were packed; the key was in it, and all I had to do was turn the key and leave when things got too bad. I was there when the *Reichstagsbrand* occurred,[15] and I remember how difficult it was for people there to understand what was going on. A friend of mine, Michael Polanyi, who was director of a division of the Kaiser Wilhelm Institute for Physical Chemistry, like many other people took a very optimistic view of the situation. They all thought that civilized Germans would not stand for anything really rough happening. The reason that I took the opposite position was based on observations of rather small and insignificant things. What I noticed was that the Germans always took a utilitarian point of view. They asked, "Well, suppose I would oppose this thinking, what good would I do? I wouldn't do very much good, I would just lose my influence. Then why should I oppose it?" You see, the moral point of view was completely absent, or very weak, and every consideration was simply consideration of what would be the predictable

[12]Prominent radiochemist, later one of the discoverers of fission.
[13]This statement was made in February, 1929.
[14]Szilard came on an immigrant visa. He left May 4, 1932.
[15]Burning of the Reichstag (House of Parliament) in Berlin, February 27, 1933.

consequence of my action. And on that basis I reached in 1931 the conclusion that Hitler would get into power, not because the forces of the Nazi revolution were so strong, but rather because I thought that there would be no resistance whatsoever.

After the Reichstag fire I went to see my friend Michael Polanyi and told him what had happened, and he looked at me and he said, "Do you really mean to say that you think that the Secretary of the Interior had anything to do with this?" and I said, "Yes, this is precisely what I mean." He just looked at me with incredulous eyes. He had at that time an offer to go to England to accept a professorship in Manchester. I very strongly urged him to take this, but he said that if he now went to Manchester he could not be productive for at least another year, because it takes that much time to install a laboratory. I said to him, "Well, how long do you think you will remain productive if you stay in Berlin?" We couldn't get together on this so I finally told him that if he must refuse this offer he should refuse it on the ground that his wife is opposed to it, because his wife could always change her mind, so that if he wanted to have the thing reconsidered he would have an out. Later on when I was in England, in the middle of '33, I was active in a committee—this was a Jewish committee incidentally—where they were concerned about finding positions for refugees from German universities. Professor Namier[16] came from Manchester and reported that Polanyi was now again interested in accepting a professorship in Manchester. He said that previously he had refused the offer extended to him on the grounds that he was suffering from rheumatism, but it appears that Hitler cured his rheumatism.

I left Germany a few days after the Reichstag fire. How quickly things move you can see from this: I took a train from Berlin to Vienna on a certain date, close to the first of April, 1933. The train was empty. The same train on the next day was overcrowded, was stopped at the frontier, the people had to get out, and everybody was interrogated by the Nazis. This just goes to show that if you want to succeed in this world you don't have to be much cleverer than other people, you just have to be one day earlier than most people. This is all that it takes.

While I was in Vienna there were the first people dismissed from German universities, just two or three; it was, however, quite clear what would happen. By pure chance I met walking in the street a colleague of mine, Dr. Jacob Marschak, who was an economist at Heidelberg and who is now [1960] a professor at Yale. He also was rather sensitive; not being German but coming from Russia he had seen revolutions and upheavals, and he went to Vienna where he had relatives because he wanted to see what was going to happen in Germany. I told him that I thought that since we were out here we might as well make up our minds what needed to be done to take up this flood of scholars and scientists who would have to leave the German universities. He said that he knew a rather wealthy economist in Vienna who might have some advice to give. His name was [Karl] Schlesinger and he had a very beautiful apartment in the Liechtensteinpalais. So we went to see him and he said, "Yes, it is quite possible that

---

[16]Sir Lewis Bernstein Namier, professor of modern history at the University of Manchester.

there will be wholesale dismissals from German universities; why don't we go and discuss this with Professor Jastrow." Professor Jastrow[17] was an economist mainly interested in the history of prices, and we went to see him—the three of us now—and Jastrow said, "Yes, yes, this is something one should seriously consider." Then he said, "You know, Sir William Beveridge is at present in Vienna. He came here to work with me on the history of prices, and perhaps we ought to talk to him." So I said, "Where is he staying?" and he said, "He's staying at the Hotel Regina." It so happened that I was staying at the Hotel Regina, so I volunteered to look up Sir William Beveridge and try to get him interested in us.

I saw Beveridge and he immediately said that at the London School of Economics he had already heard about dismissals and he was already taking steps to take on one of those dismissed, and he was all in favor of doing something in England to receive those who had had to leave German universities. So I phoned Schlesinger and suggested that he invite Beveridge to dinner. Schlesinger said no, he wouldn't invite him to dinner because Englishmen, if you invite them to dinner, get very conceited. However, he would invite him to tea. So we had tea, and in this brief get-together with Schlesinger and Marschak[18] and Beveridge, it was agreed that Beveridge, when he got back to England, and when he got the most important things which he had on the docket out of the way, would try to form a committee which would set itself the task of finding places for those who had to leave German universities. He suggested that I come to London and that I occasionally prod him on this, and if I prodded him long enough and frequently enough he thought he would do it. Soon thereafter he left, and soon after he left, I left and went to London. When I came to London I phoned Beveridge who said that his schedule had changed and he found that he was free and that he could take up this job at once. This is the history of the birth of the so-called Academic Assistance Council in England.

The English adopted a policy of mainly helping the younger people; they did not demand that somebody should have an established name or position in order to find a position for him in England, quite in contrast to American organizations. In addition to the Academic Assistance Council there was a Jewish committee functioning. They raised funds privately and they found positions for people and provided them with fellowships for one or two years. The two committees worked very closely together, and in a comparatively short time practically everybody who came to England had a position, except me.[19] (*Docs. 2–4*)

When I got to England, and after I no longer had to function in connection with placing the scholars and scientists who had left the German universities—when this

---

[17]Ignaz Jastrow, professor of political science at the University of Berlin.
[18]Marschak later said that he did not attend this tea, although he was aware of it. Marschak to Edward Shils, October 4, 1964.
[19]Szilard had not told them he needed a job. For this work see Laura Fermi, *Illustrious Immigrants. The Intellectual Migration from Europe 1930–1941* (Chicago: University of Chicago Press, 1971), pp. 63–64, 66; Norman Bentwich, *The Refugees from Germany, April 1933 to December 1935* (London: Allen & Unwin, 1936), chapter 9.

was more or less organized and there was no need for me to do anything further about that—I was thinking about what I should do, and I was strongly tempted to go into biology. I went to see A. V. Hill[20] and told him about this. A. V. Hill himself had been a physicist and became a very successful biologist, and he thought this was quite a good idea. He said, "Why don't we do it this way. I'll get you a position as a demonstrator in physiology, and then twenty-four hours before you demonstrate, you read up these things, and then you should have no difficulty in demonstrating them the next day. In this way, by teaching physiology, you would learn physiology, and it's a good place to begin."

Now I must tell you why I did not make this switch at that time. In fact, I made the switch to biology in 1946. In 1932, while I was still in Berlin, I read a book by H. G. Wells. It was called *The World Set Free*.[21] This book was written in 1913, one year before the World War, and in it H. G. Wells describes the discovery of artificial radioactivity and puts it in the year of 1933, the year in which it actually occurred. He then proceeds to describe the liberation of atomic energy on a large scale for industrial purposes, the development of atomic bombs, and a world war which was apparently fought by an alliance of England, France, and perhaps including America, against Germany and Austria, the powers located in the central part of Europe. He places this war in the year 1956, and in this war the major cities of the world are all destroyed by atomic bombs. Up to this point the book is exceedingly vivid and realistic. From this point on the book gets to be a little, shall I say, utopian. With the world in shambles, a conference is called in Brissago in Italy, in which a world government is set up.

This book made a very great impression on me, but I didn't regard it as anything but fiction. It didn't start me thinking of whether or not such things could in fact happen. I had not been working in nuclear physics up to that time.

This really doesn't belong here, but I will nevertheless tell you of a curious conversation which I had, also in 1932, in Berlin. The conversation was with a very interesting man named Otto Mandl, who was an Austrian and who became a wealthy timber merchant in England, and whose main claim to fame was that he had discovered H. G. Wells at a time when none of his works had been translated into German. He went to H. G. Wells and acquired the exclusive right to publish his works in German, and this is how H. G. Wells became known on the Continent. Something went wrong with his timber business in London, and in 1932 he found himself again in Berlin. I had met him in London and I met him again in Berlin and there ensued a memorable conversation.[22] Otto Mandl said that now he really thought he knew what it would take to save mankind from a series of ever-recurring wars that could destroy it. He said that Man has a heroic streak in himself. Man is not satisfied with a happy idyllic life: he has the need to fight and to encounter danger. And he concluded that what mankind must do to save itself is to launch an enterprise aimed at leaving the earth. On this

---

[20]Archibald Vivian Hill, FRS, professor of physiology, University of London.
[21]*The World Set Free: A Story of Mankind* (London: Macmillan, 1914).
[22]Otto Mandl's widow, the pianist Lili Kraus, recalls that these discussions took place as described.

task he thought the energies of mankind could be concentrated and the need for heroism could be satisfied. I remember very well my own reaction. I told him that this was somewhat new to me, and that I really didn't know whether I would agree with him. The only thing I could say was this: that if I came to the conclusion that this was what mankind needed, if I wanted to contribute something to save mankind, then I would probably go into nuclear physics, because only through the liberation of atomic energy could we obtain the means which would enable man not only to leave the earth but to leave the solar system.

I was no longer thinking about this conversation, or about H. G. Wells' book either, until I found myself in London about the time of the British Association [meeting] in September, 1933. I read in the newspapers a speech by Lord Rutherford. He was quoted as saying that he who talks about the liberation of atomic energy on an industrial scale is talking moonshine.[23] This sort of set me pondering as I was walking the streets of London, and I remember that I stopped for a red light at the intersection of Southampton Row. As I was waiting for the light to change and as the light changed to green and I crossed the street, it suddenly occurred to me that if we could find an element which is split by neutrons and which would emit *two* neutrons when it absorbed *one* neutron, such an element, if assembled in sufficiently large mass, could sustain a nuclear chain reaction. I didn't see at the moment just how one would go about finding such an element, or what experiments would be needed, but the idea never left me. [2] In certain circumstances it might become possible to set up a nuclear chain reaction, liberate energy on an industrial scale, and construct atomic bombs. The thought that this might be in fact possible became a sort of obsession with me.[*1*]

Soon thereafter, when the discovery of artificial radioactivity by Joliot and Mme. Joliot was announced, I suddenly saw that the tools were at hand to explore the possibility of such a chain reaction.[24] I talked to a number of people about this. I remember that I mentioned it to G. P. Thomson[25] and to Blackett,[26] but I couldn't evoke any enthusiasm. [2] I had a little money saved up, enough perhaps to live for a year in the style in which I was accustomed to live, and therefore I was in no particular hurry to look for a job. I moved to the Strand Palace Hotel and started to dream about the possibilities which had been opened up by the recent discoveries in physics. [6] (*Doc. 6*)

I had one candidate for an element which might be unstable in this sense of splitting

---

[23]A summary of the speech by Ernest Rutherford (the famous Cambridge physicist), delivered at the meeting of the British Association for the Advancement of Science, Leicester, September 11, 1933, and published in *Nature, 132*: 432–433 (September 16, 1933), contains the sentence: "One timely word of warning was issued to those who look for sources of power in atomic transmutations—such expectations are the merest moonshine."

[24]Irène Curie and Frédéric Joliot, *Comptes Rendus de l'Académie des Sciences (Paris), 198*: 254–256 (1934). At this time the Joliots were at the Radium Institute, Paris. They noted that bombardment with alpha-particles would make some elements radioactive. Szilard realized that neutron bombardment would also do this.

[25]George Paget Thomson, professor of physics at the Imperial College of Science, London.

[26]P. M. S. Blackett, professor of physics at the University of London.

off neutrons when it disintegrates, and that was beryllium. The reason that I suspected beryllium of being a potential candidate for sustaining a chain reaction was the fact that the mass of beryllium was such that it could have disintegrated into two alpha particles and a neutron. It was not clear why it didn't disintegrate spontaneously, since the mass was large enough to do that; but it was conceivable that it had to be tickled by a neutron which would hit the beryllium nucleus in order to trigger such a disintegration. I told Blackett that what one ought to do would be to get a large mass of beryllium, large enough to be able to notice whether it could sustain a chain reaction. Beryllium was very expensive at the time, almost not obtainable, and I remember Blackett's reaction was, "Look, you will have no luck with such fantastic ideas in England. Yes, perhaps in Russia. If a Russian physicist went to the government and says, 'We must make a chain reaction,' they would give him all the money and facilities which he would need. But you won't get it in England." As it turned out later, beryllium cannot sustain a chain reaction and it is in fact stable. What was wrong was that a published mass of helium was wrong. This was later on discovered by Bethe,[27] and it was a very important discovery for all of us, because we did not know where to begin to do nuclear physics if there could be an element which could disintegrate but doesn't.

When I gave up the beryllium, I did not give up the thought that there might be another element which could sustain a chain reaction. In the spring of 1934 I had applied for a patent which described the laws governing such a chain reaction. This was the first time, I think, that the concept of critical mass was developed and that a chain reaction was seriously discussed. Knowing what this would mean—and I knew it because I had read H. G. Wells—I did not want this patent to become public. The only way to keep it from becoming public was to assign it to the government. So I assigned this patent to the British Admiralty.[28]

At some point I decided that the reasonable thing to do would be to investigate systematically all the elements. There were ninety-two of them. But of course this is a rather boring task, so I thought that I would get some money, have some apparatus built, and then hire somebody who would just sit down and go through one element after the other. The trouble was that none of the physicists had any enthusiasm for this idea of a chain reaction. So I thought, there is after all something called "chain reaction" in chemistry. It doesn't resemble a nuclear chain reaction, but still it's a chain reaction.[29] So I thought I would talk to a chemist, and I went to see Professor

---

[27] Hans Bethe, physicist at Cornell; see Bethe, *Physical Review*, 47:633–634 (1935).
[28] British patents 440,023 (applied for March 12, 1934); 630,726 (applied for June 28, 1934), in *Scientific Papers*, pp. 622–651. See Szilard to C. S. Wright, February 26, 1936, ibid., pp. 733–734. Szilard turned to the Admiralty only after the War Office had turned him down with the statement that "there appears to be no reason to keep the specification secret so far as the War Department is concerned." J. Coombes, Director of Artillery, to Claremont Haynes & Co. (Szilard's patent attorneys), October 8, 1935, Szilard Papers.
[29] In a chemical chain reaction the reaction is spread by the release of energy; in a nuclear chain reaction by the release of particles (neutrons).

Chaim Weizmann, who was a renowned chemist and a Zionist leader. I had met him on one occasion or another. And Weizmann listened and Weizmann understood what I told him. He said, "How much money do you need?" I said that I thought that £2,000 would be enough to do this; it would have been at that time about $10,000. So Weizmann thought that he would try to get this money. I didn't hear from him for several weeks, but then I ran into Michael Polanyi, who by that time had arrived in Manchester and was head of the chemistry department there. Polanyi told me that Weizmann had come to talk to him about my ideas for the possibility of a chain reaction, and he wanted Polanyi's advice on whether he should get me this money. Polanyi thought that this experiment ought to be done. Then again I didn't hear anything. As a matter of fact I didn't see Weizmann again until the late fall of '45, after Hiroshima. I was at that time in Washington and I ran into him in the Wardman-Park Hotel. He seemed to be terribly happy to see me, and he said, "Do you remember when you came to see me in London?" I said, "Yes." He said, "And do you remember what you wanted me to do?' I said, "Yes." And he said, "Well, maybe you won't believe me, but I tried to get those £2,000 and found that I couldn't."[2] (*Docs. 7–15*)

I have often thought since that time that if Weizmann hadn't failed me, almost certainly Germany would have won the last war. For even though I was fully aware of the implications and determined to try what I could to keep the experiment a secret, it is almost certain that in prewar England it would not have been possible to keep such a discovery a secret, and that neither in England nor in America would this development have been pushed with the determination in which it would have been pushed in Germany, which in 1935 was fully determined to rearm and go to war.[30] [7]

Because of these thoughts about the possibility of the chain reaction and because of the discovery of artificial radioactivity, physics became too exciting for me to leave it. So I decided not to go into biology as yet, but to play around a little bit with physics, and I spent several months in the spring at the Strand Palace Hotel, doing nothing but dreaming about experiments which one could do, utilizing this marvelous tool of artificial radioactivity which Joliot had discovered. I didn't do anything; I just thought about these things. I remember that I went into my bath—I didn't have a private bath, but there was a bath in the corridor in the Strand Palace Hotel—around nine o'clock in the morning. There is no place as good to think as a bathtub. I would just soak there

---

[30] A memorandum in the Szilard Papers, probably written about July, 1934, "On Nuclear Chain Reactions and their Bearing on the Question of Power Production," shows that the experiments Szilard planned were essentially of the types that eventually demonstrated neutron multiplication in uranium. With uncanny prescience he saw the possibility of observing neutrons produced in a chain reaction, either by noting their action on other elements sensitive only to fast neutrons, or by noting their distribution in the reacting mass; he also foresaw that small bits of material could be used to map out this distribution through the radioactivity induced in them by neutrons. These experimental techniques were independently invented and developed by Enrico Fermi, Otto Frisch, and others in the mid- and late-1930s. Other documents in the Szilard Papers (e.g., *Doc. 13*) show that uranium was one of the three or four elements whose properties puzzled Szilard and which therefore he found likely candidates for a chain reaction. Had his experiments been funded and pursued he probably would have discovered uranium chain reactions well before 1939.

and think, and around twelve o'clock the maid would knock and say, "Are you all right, sir?" Then I usually got out and made a few notes, dictated a few memoranda. I played around this way, doing nothing, until summer came around. At that time I thought that one ought to try to learn something about beryllium. I thought that if beryllium really is so easy to split, the gamma rays of radium should split it and it should split off neutrons.

I had casually met the director of the physics department of St. Bartholomew's Hospital, so I dropped in for a visit and asked him whether in the summertime, when everybody was away, I could not use the radium, which was not much in use in summer, for experiments of this sort. And he said, yes, I could do this; but since I was not on the staff of the hospital I should team up with somebody on his staff. There was a very nice young Englishman, Mr. Chalmers,[31] who was game, and so we teamed up and for the next two months we did experiments. It turned out that in fact beryllium splits off neutrons when exposed to the gamma rays of radium. This later on became really very important, because these neutrons are slow neutrons. Therefore if they disintegrate elements like uranium—of course we didn't know that until after the discovery of Hahn[32]—and if in that process neutrons come off, which are fast, you can distinguish them from neutrons of the source, which are slow.

We did essentially two experiments. We demonstrated that beryllium emits neutrons if exposed to the gamma rays of radium, and we demonstrated something else, which is called the Szilard-Chalmers effect.[33] These experiments established me as a nuclear physicist, not in the eyes of Cambridge, but in the eyes of Oxford. [2] I had never done work in nuclear physics before, but Oxford considered me an expert in nuclear physics. Cambridge, which was a stronghold of Rutherford, would never have made that mistake. For them I was just an upstart who might make all sorts of observations, but these observations could not be regarded as discoveries until they had been repeated at Cambridge and confirmed. [7]

There was an international conference on nuclear physics in London in September where these two discoveries were discussed by the participants,[34] and so I got very favorable notice. This led within six months to an offer of a fellowship at Oxford. However, I didn't get this offer until I had left England and come to America, where I didn't have a position but I had some sort of a fellowship. When I received this offer from Oxford, I had the choice of either keeping on this fellowship in America or returning to Oxford. I sat down and wrote a letter to Michael Polanyi in which I described my choice between these two alternatives, and this is what I wrote him: that I would accept the fellowship at Oxford and go to England, and I would stay in

[31]T. A. Chalmers, a member of the physics department, Medical College, St. Bartholomew's Hospital, London.
[32]Otto Hahn and Fritz Strassmann, *Naturwissenschaften, 27*:11–15 (January 6, 1939).
[33]Szilard and T. A. Chalmers, *Nature, 134*:494–495, 462–463 (1934), in *Scientific Papers*, pp. 143–145.
[34]*International Conference on Physics, London, 1934: Volume I, Nuclear Physics* (Cambridge, England: Cambridge University Press, 1935), pp. 88–89.

England until one year before the war, at which time I would shift my residence to New York City. That was very funny, because how can anyone say what he will do one year *before* the war? So the letter was passed around and a few people commented on it when I finally turned up in England.

However, this is precisely what I did. [2] When the German troops moved into the Rhineland and England advised France against invoking the Locarno pact [March 1936], I knew that there would be war in Europe. [1] In 1937 I decided that the time had come for me to change my fulltime fellowship at Oxford to one which permitted me to spend six months out of the year in America. And on the basis of that arrangement (I had to take a cut in salary, of course—I had to go on half pay, so my total income amounted to $1,000 a year) I came over to America [on January 2, 1938]. And I did nothing but loaf. I didn't try to look for a position; I just thought I would wait and see. Then came the Munich crisis. I was visiting Goldhaber[35] in Urbana, Illinois, at the time that the Munich crisis broke [August-September, 1938]. I spent a week listening to the radio giving news about Munich, and when it was all over I sat down and wrote a letter to [F. A.] Lindemann, later Lord Cherwell, who was director of the Clarendon Laboratory [at Oxford] where I was employed. The letter said that I was now quite convinced that there would be war, and therefore there would be little point in my returning to England unless they would want to use me for war work. If, as a foreigner, I would not be useful for war work, I would not want to return to England but rather stay in America. And so I resigned at Oxford and stayed here. [2] (*Docs. 16–18*)

[35]Maurice Goldhaber, assistant professor of physics, University of Illinois.

Documents through January 1939

"In the mid-twenties in Germany," Szilard wrote, "I became quite convinced that the parliamentary form of democracy would not have a very long life in Germany, but I thought that it might survive one or two generations. . . . It seemed to me therefore at that time that it might make good sense to create an organization in Germany which might fulfill a useful function within the framework of the then existing political system, and which in addition [would] be capable of growing and ultimately might stand ready to exercise the functions of government if and when the parliamentary system in Germany collapsed, one or two generations hence.

"In devising such an organization, which I called the Bund,[1] *I drew upon the experience provided by the history of the so-called* Jugendbewegung [*youth movement*], *a spiritual and moral movement among the youth of Germany, which originated and culminated before the First World War. This movement had represented what was best in Germany and had a profound and on the whole beneficial effect, extending for about one generation beyond its cessation."*[2]

In 1930 Szilard tried to set the plan in motion in England and Germany. Einstein, answering a question from a possible supporter, wrote that "Szilard has assembled a circle of excellent young people, mostly physicists, who are in sympathy with his ideas. But as yet there is no organization of any kind." He added that he considered Szilard "a fine, intelligent man, who is ordinarily not given to illusions. Perhaps, like many such people, he is inclined to overestimate the significance of reason in human affairs."[3] *The early 1930s proved a poor time for ventures like Szilard's "Bund."*

The scheme turned out to be important less as a practical effort than as a crystallization of the private dreams which lay behind many of Szilard's later plans. Parts of this strange document are included here because it seems to be the earliest of his innumerable schemes "to save the world." While this design remained a dream, subsequent efforts, such as the "Pugwash Movement" and the "Angels Project," were more realistic and culminated in 1962 in the establishment of "The Council for Abolishing War." The Council continues to this date, fifteen years later, as "The Council for a Livable World," a potent political organization in Washington, D.C., run largely by the team which Szilard selected and brought together. These efforts are documented in detail in a draft

---

[1] An untranslatable German word, in this context connoting a closely bonded alliance of like-minded people.
[2] From a memorandum dated January 26, 1949, prepared for the Ford Foundation but probably not sent. The plan may have been inspired not only by the youth movement but also by H. G. Wells' *The Open Conspiracy: Blue Prints for a World Revolution* (London: V. Gollancz, 1928), urging the creation of groups which would work toward a cooperative world commonweal infused by a new religious spirit. Szilard, who met Wells in 1929, admired his work and used the "open conspiracy" term in later years. But Szilard may have conceived the "Bund" independently and later responded enthusiastically to Wells' work because the ideas were so consonant with his own.
[3] Einstein to H. N. Brailsford, April 24, 1930, Einstein Archives, Princeton; see Otto Nathan and Heinz Norden, eds., *Einstein on Peace* (N.Y.: Simon & Schuster, 1960), pp. 103–104.

manuscript presenting those social and political papers written by Szilard after 1946 and yet to be published.

He retained some incomplete drafts, with many revisions, of his plan. One section was translated into English and partly corrected by Szilard[4]; *the rest of the following is our translation of sections of what seems to be the last version.*

Document 1

[Draft of a proposal for a new organization called "Der Bund," about 1930]

In the "Society of the Friends of the Bund" those persons should come together who wish to further the creation of a closely knit group of people whose inner bond is pervaded by a religious and scientific spirit. The external frame into which such a group should grow would be the "Bund" itself.

The Bund would play a many-faceted role. It could accomplish much very soon, much some time hence, and some things at best in the distant future. Only many years' experience can show whether and how far the hopes (some very far-reaching) are still justified. But nobody should renounce even the boldest hopes before human nature has been given every opportunity to demonstrate its limits.

We will begin here with those possibilities which lie in the remotest future and put the political perspectives at the top of our consideration. Thus the reader, knowing our original motives and understanding the spirit in which we operate, can easily complete the sketchily drafted work plan and avoid misunderstandings. We trust that these general observations will not distract the reader from the concrete work to be done.

Political Views and Motives

One must recognize that the great successes of the past, right up to the present, were won under the rule of laissez-faire. The development of Western civilization was not brought about by humanity's yearning for civilization but rather arose in a more mechanical way. The life-styles predominating in civilized countries today are no doubt closely associated with the ruling economic system and largely determined by the mechanics of those forces which appear as the moving forces behind this economic system. At this time there exists hardly anything resembling the formation of a community purpose,[5] and in any case very little depends on the ideas, which may exist more or less clearly defined, in the heads of the current official rulers.

Laissez-faire is to a certain degree built into the parliamentary-democratic system;

---

[4]The sentences beginning, "Now we shall say a few words about the Bund itself . . ." and ending ". . . some kind of task which one of the constituted groups of the Bund have taken over."
[5]*Willensbildung in der Gemeinschaft.*

at least the political parties working against one another and the brief periods in office of the successive governments work in this direction.

Perhaps those who see herein an advantage to the parliamentary system are right. For as long as the formation of a community purpose continues in such a primitive way as is the case today, as long as today's press forms the stage for the battle of public opinion, as long as this battle is fought in the contemporary manner and the deciding forum is the misinformed newspaper reader, laissez-faire does not seem to be the worst choice.

It is improbable, however, that things will remain this way forever. Already today the role of the permanent civil servant grows in importance daily in parliamentary democracies, and states are beginning to turn away from laissez-faire to a considerable degree despite the parliamentary form of government. Herein lies no small danger, for in today's bureaucracy there is no structure whatsoever that would offer a guarantee of a well-thought-out, large-scale definition of aims.

In this regard we are also doubtful that the forms of government in existence in Italy[6] and Russia today—should they retain their present structure—will last for any great length of time. For it does not seem unimportant to us that the formation of a community purpose should take place in the midst of a free struggle for some sort of public opinion, even though it may be merely before a forum which consists numerically of only a small part of the total population. Beyond this it also appears that a democratic situation is indispensable, in the sense that what government does must not run counter to the public opinion of the broadest stratum. To remain democratic in this sense seems almost a permanent requirement, while we may perhaps have to regard parliamentary democracy as an outmoded form of government in the not-too-distant future. In fact it would be outmoded the moment one presumed in good conscience definitely to give up laissez-faire through the progressive development of public life. Then one would have to deal with the question of whether there is something better to substitute for it, and it would be rather risky if one were to make a decision based on theoretical speculations.

Utopia

If we possessed a magical spell with which to recognize the "best" individuals of the rising generation at an early age (and we would be wise to refuse to specify such a magical charm in advance) then we would be able to train them to independent thinking, and through education in close association we could create a spiritual leadership class with inner cohesion which would renew itself on its own. Members of this class must not of course be entitled to a higher standard of living nor to personal glory. Rather selected people should demonstrate their devotion through particular burdens which they take upon themselves and through a life of service. In this sense one would be tempted to speak of an "order"; yet entry into this class and advancement into

---

[6]Mussolini's Fascist dictatorship was established in Italy between 1922 and 1924.

higher spheres of action within it should not be modeled after religious orders. For there would be a great danger that the ruling opinion formulated inside the leading class would be retained for a very long time. The same would be true of an institution based on the scientific academies, replenishing itself through inside elections, namely, that as a consequence of the attempt to preserve the ruling opinion, new members might be selected from a "political" standpoint. Only if some kind of public opinion is continuously formed anew within this class of leaders by a reasonably fought contest can there exist the prospect of some effectiveness which would keep up with the evolution of ideas over longer periods of time.

If such a group, profoundly cohesive in spirit, were to exist, then presumably it would exercise a potent influence on the shaping of public affairs even without any particular inner structure and without any constitutionally determined rights. There are sufficient examples from the past where certain groups of people or institutions exercised a decisive influence without occupying a position specified by the constitution.

It would also be conceivable that such a leading group would take over a more direct influence on public affairs as part of the political system, next to government and parliament, or in the place of government and parliament. If this is to be possible with some hope of success, then care must be taken that the prevailing opinion of the leading group be safely and freely communicated to a wide public so that the two may never diverge to a significant degree.

The choice of recruits and advancement of individuals to ever higher spheres of activity would have to be thought over and tested with great care. Before thinking of finding legal bases for an institution one should, we believe, wait until it has adequately demonstrated its suitability for the work of reconstruction.

The Bund

Now we shall say a few words about the Bund itself and shall give an outline of it as it would appear in the course of time after having reached its full development. In reality this fully developed form can only be reached gradually.

The First Step

New members will be drawn from [the "best" of] the top form of the Secondary schools; this will be boys and girls of 18 to 19 years of age. We will discuss later what is meant by the "best" and how they will be selected.

What we want are boys and girls who have the scientific mind and a religious spirit. Let us assume that we possess some magic means of making the right choice. Those selected will be the "Junior" members of the Bund. It is essential that this selection should be satisfactory as regards the character of the individual.

Schoolmasters, the Bund or the Friends of the Bund should hardly be entrusted with the choice as they might be influenced by political considerations.

To become a Junior member should be considered a distinction. Opportunities will be given to the Juniors to prove their devotion by doing some definite piece of work [and they] could choose fairly freely the kind of work which suits them best. The service may consist of carrying out some kind of task which one of the constituted groups of the Bund have taken over. Or they can perform their "service" by undertaking to acquire certain knowledge, or skills, within a certain period of time. For example, the learning of a language of a neighboring country within the year could be chosen as a real "service."

The Junior members will be given every opportunity to get to know each other. Clubrooms with reading and lecture rooms will be at their disposal—places where young people of different ages can come together. Here, under the guidance of the Senior members, their interest in public affairs will be stimulated at an early age.

Here they will learn to think for themselves and do so in areas where for most people passions and emotions prevent clear thinking. They shall also be taught *not* to take sides on any problems for which they have not arrived at a judgment of their own.

To keep them open-minded as long as possible, they should not be allowed to join, even as a formality, any political party or philosophical movement[7] before they have reached the age of 30 years. On the other hand they should have ample opportunity to become familiar with all of the political movements, by coming into contact with the finest representatives of the various movements.

On the other hand, the Junior members shall be in no way isolated from their peers in the total population. It is not yet clear to what extent it will be possible to maintain this close contact between the Junior members and those of their comrades who have already left high school. Above all, however, the Juniors should maintain the closest contact with the younger elements (from the ages 15 to 19) who are still in school. The club-rooms should be one way of providing them with the opportunity of frequently getting together.

The Second Step

Junior members should be concentrated at relatively few universities and should live in close contact with one another. Special seminars and workshops, those arranged by the university and those arranged by the Bund or the Friends of the Bund under the direction of a faculty member of the university, should bring into contact 30 to 40 Junior members. Students who do not belong to the Bund, but who are recognized as outstanding at the university, could be invited to join these workshops and be able to participate.[8]

These workshops will enable the Junior members to get acquainted with each others' thinking. Anyone who has led seminars for several semesters knows that it does not

---

[7] *weltanschauliche Partei.*
[8] Such workshops and other youth organizations were very popular at German universities in the 1920s.

take long for the participants to know exactly who amongst them knows something and who does not. After being together for several years the various groups of Juniors now choose from amongst themselves the "best" and these become the Senior members of the Bund. *We expect that this second selection will prove satisfactory at least as far as the intellect of those chosen is concerned.* (Those participants of the workshops and seminars who do not belong to the Bund as Juniors but who have been drawn in from outside have no active voting rights but can be elected as Seniors. They have, then, the same rights as their comrades of the Bund, including a passive electoral right, but without ever having been granted an active voice in the Bund.)

The Junior members shall be completely free to study at the university the subjects which appeal to them most. One may assume that there will be a general interest in certain subjects, at least amongst those interested in public affairs. Emphasis will be on workshops and seminars in these subjects [public affairs] in addition to those which test the students' ability to think.

INTERNATIONAL EXCHANGE

Persons interested in exercising some influence on public affairs today can hardly do so without understanding the mentalities of other nations.

The Seniors, therefore, shall be given the opportunity of spending one to two years, after leaving the university, as the guests of the Bund in a neighbouring country where they can either continue their studies or do some practical work. For this purpose they must have become proficient in the appropriate language during their university years. For those Seniors who wish to devote themselves later to public affairs, a period of residence in a foreign country is obligatory.

The Third Step

After graduating from the University, or on their return from abroad, many Seniors will wish to devote themselves to their professions or to their families and will wish only to maintain a more or less loose contact with the Bund. To others, however, who feel the need for a contact with the Bund, closer than just from the intellectual or personal friendship point of view, the opportunity is presented to enter the "Order" of the Bund. The Order imposes on its members a life of sacrifice and service. The sacrifice shall be so severe that this path will only be followed by those who are imbued with the desired spirit, but will not be such as to hinder the man's efficiency. The Order should also comprise different degrees of severity.

The members of the Order shall have, in principle, full freedom to work in those professions and places which they consider suitable. They must hand over to the Order all monies they earn above the base minimum necessary for their existence.

The funds thus acquired by the Order shall serve, primarily, to assure the donor of this minimum in case of future need; and secondly, to relieve other members of the

Order from the worry of having to adopt a career. The required minimum may vary according to the different degrees of severity imposed by the Order and the number of children shall be taken into account. For a part (about a third?) of the members of the Order, celibacy may perhaps be prescribed.

The Bund will make every effort to ensure for the members of the Order a suitable sphere of action. Many will want to work in those fields which are more or less related to public affairs. They will supposedly soon find their place not only in university seminars, in industrial enterprises, in state or community services, but also in important administrative positions of individual industrial establishments.

The members of the Order shall remain in very close touch with one another and, brought together in committees in which outsiders may also participate, work on solving problems that are pertinent and of interest to society. These committees, whose duty it is to carry out certain well-defined tasks, constitute the cells that constitute the Order. Such a cell must not contain more people than can get to know each other personally and very well (about 30 to 40). The only way leading to higher and higher responsibility for someone within the Order is the designation as the "best" by the members of his cell who know him well. A certain degree of freedom between the individual cells shall eliminate the disadvantages of such methods of selection.

The careful selection, the good education, and the great devotion of the members of the Order assure them of a certain advantage over the others. In addition, they are inclined to promote each other in contrast to persons who think of themselves in the middle of a struggle of everyone against everyone else. And finally a deciding factor is also the circumstance that they are permitted to concentrate their attention more on a factual task than on the advancement of their own career in their most active years, between the ages of 25 and 35. Thus it seems quite feasible that by the weight of the personality of the individual members and by the cohesiveness of this group, the Order might represent some form of structure in public life, which would leave an imprint on the whole sprutual life of the community. It is probable that subsequently other patterns will arise spontaneously, which will safeguard the transmission of public opinion—forever renewed within the Order in difficult strife—to the general population.[9]

---

[9]In other fragments Szilard developed this idea further. He translated these in his memorandum of 1949 (see note 2 of the introduction preceding this document) as follows: "The Order was not supposed to be something like a political party . . . but rather it was supposed to represent the state. The same political differences of opinion existing in the state would also perforce exist within the Bund. These would have to be fought out within the Bund and in particular these would have to be fought out on the top level, represented by a group composed of perhaps 40–50 people. Because of the method of selection [and education] . . . there would be a good chance that decisions at the top level would be reached by fair majorities.

". . . It appeared likely that gradually something like a political platform would be formulated within the Bund that would have the approval of a substantial majority of its members. Because of the position which the members of the Bund were supposed to occupy in public life, because they were expected to occupy key positions in education and other fields of communication, and also because of the close contact which the Bund was supposed to maintain by means of its junior clubs with the incoming generation, it was assumed that the country as a whole would be responsive to the leadership of the Bund, and that there was not much danger that the policies which gradually became victorious within the Bund would not command the strong support of the people."

Whether one should ever give the Order an opportunity to exercise a more direct influence on public affairs is a question that can be postponed until experience has shown how the Order stands the test of time and, above all, to what degree it has succeeded in remaining closely bound to the general public.

Development

Regarding the practical development, the first consideration will be the selection of the best people between 18 to 19 years old. From the outset we would reject a selection by the teachers alone, based on tests or exams, or a selection solely through the Bund itself or even through the "Society of the Friends of the Bund." It remains to be seen whether anything can be achieved through a combination of several methods of selection, when each one would be unsuitable if applied alone. Perhaps we shall have to try out a variety of methods.

The first thing we should try is to find out, by interviewing the children themselves, which children are looked upon as personalities by their peers. It is certain that children who have spent many years together in the same grade know each other very well by the time they are 18 to 19 years old. Nevertheless, only experience will show what type of human material this method supplies under varying circumstances.

To begin with, we should select the three best children from a class of about 30 to 40 by questioning the children themselves and these three should select a fourth from the class. The children selected in this manner are the Juniors of the Bund, and the gradual development should follow in such a way that the organization begins with a few schools and expands to include more and more schools.

The initial phase should proceed in at least two different countries simultaneously so that international exchange can begin at once. We have in mind first of all two of the three countries, Germany, England, and France, and for the selection the only determining factor should be the least resistance to the practical realization.

For the time being the Order of the Bund should not be brought into being so that available intellectual and financial resources can be concentrated on the Juniors and their University studies. Of primary importance in financing are the building of clubrooms for the Juniors and support for university studies for those Juniors who have insufficient funds.

In the near future the Friends of the Bund will be obliged to make it possible to study at universities for those Juniors who would otherwise not have the means to do so. In some countries this task will present the greatest burden to the budget and will be in the foreground for some time, in addition to the maintenance of the premises of the club and the international exchange of the Seniors.

The necessary monies should be obtained in part from private sources, in part through the State, and in part from the students themselves. As a matter of principle the monies collected from private sources should not be provided unconditionally but should require matching funds from the State or the students themselves.

The individual classes which send Juniors to the Bund should be invited to cooperate with the Bund by having the students pledge a small fraction of *their* income earned after leaving school on a continuing basis for the Bund. Classes whose students refuse to participate in this should be regarded as inferior and for the time being Juniors should not be selected from these classes. The sacrifice which the children take upon themselves in this manner should strengthen the bond that ties them to the Bund. . . .

The Role of the Friends of the Bund

The role of this organization will at first be very important and can then gradually decrease, in the measure in which the Bund itself is built up and finally is in a position to administer itself.

The Friends of the Bund must see that the leaders who are close to the various political movements will have equal opportunity to present their ideas before the Juniors. The persons who are close to the Friends of the Bund shall also have an opportunity to present their own views before the Juniors but this must not take more time than the presentations of the various outside views.

The effect of the Friends of the Bund, if indeed it becomes an effect, must be analogous to the effect of a seed crystal thrown into a supersaturated solution; if the solution is indeed supersaturated and the seed crystal the appropriate one, then very soon the mass crystallizes by itself. One has to trust that children who are able to think, who have been educated to independent thinking, and who seriously go about it will discern the truth from the confusion of conflicting opinions, relying on their own judgement. The only weapon that the Friends of the Bund may apply in their struggle for the Bund is the effect of the personality and the clear and consistent concept that they can present versus the confusion of ideas existing today.

The following group of letters indicates Szilard's work and plans for rescuing refugee scientists and his ideas for an international board of scientists:

Document 2

Imperial Hotel, Russell Square
May 4, 1933

Sir William Beveridge
London School of Economics
Aldwych, W.C.1.

Dear Sir William:

We talked last time about the fact that some Jewish groups may wish to raise money for purposes which are different from ours. There are two such plans: A. to raise money for the Palestine University, B. to raise money for an emigrants' university to be founded somewhere in Europe. I saw to-day Sir Philip Hartog,[10] and spent the day attending the meeting of the Jewish committee of which he is the chairman.

Sir Philip Hartog agrees that alternative B. is not desirable. I informed Sir Philip that there is a vague hope of getting Oxford, Cambridge and London universities to take the matter into their own hands in England, and that they may take some steps to raise funds. He seemed to be quite willing to prevent anything that would counteract or interfere with such an action on the part of the universities. As he will lunch with you on Friday, I need not go into further details.

Dr. Weizmann's secretary wrote me a letter saying that Dr. W. will see me one of these days.

I do not think I can do anything to persuade Weizmann to divert funds from Palestine University for our purpose, but I shall inform him of the consensus of opinion as far as our project is concerned, among those German professors whom I happened to meet in London during the last few days.

Enclosed I am sending you a copy of a letter which I had to-day from Belgium. As you see from this letter, there are many groups everywhere already in existence, and there is a definite need to coordinate these groups.

A possible way of doing this would be to have an international board of some twenty scientists and scholars, including some German professors who are remaining in office, for instance Planck and Hilbert.[11] This board would not need to meet but each of the members would assume responsibility for selecting fellows in his field.

If we had such a central body of prominent scientists and scholars, the fellowships which would be granted through this body would not earmark those scientists who get them, even if some of the groups who raised the money would be antagonistic to the German regime in one way or another. The statutes of the board would merely have to state the willingness of the board to give advice if required concerning research fellowships, and the cooperation with the individual groups could be as loose or as close as would prove suitable.

I have talked over this point with several people, also with professor Laski,[12] and I would very much like to have your opinion on it before I inform accordingly our Belgian and American friends. Professor Laski and myself agreed that it would not be wise to appoint such a board at the present time, but that we could secure right now the consent of those who would be officially asked later.

I shall possibly try and see you some time to-morrow morning, and will get in touch with Mrs. Turin for this purpose.

Yours sincerely,
Leo Szilard

[10]Sir Philip J. Hartog, chemist and educator. See Szilard to "D.," *Doc. 3.*
[11]David Hilbert, mathematician in Göttingen.
[12]Probably Harold J. Laski, professor of political science at the University of London.

PS. I happened to meet Professor Donnan (physical chemistry, U.C.).[13] He has already been in touch with Lord Melchett[14] who promised him his help, and it may be of some use that you meet him sooner or later. I also happened to see Sir John Russell[15] at Harpenden.

Document 3

Imperial Hotel
London, W.C.1.
May 7, 1933

Dear Dr. D.,[16]

Enclosed you will find an outline of the work before us.

I would like also to inform you of my part of the work. I got in touch in Vienna with Sir William Beveridge, the Director of the London School of Economics, who happened to be there, and I discussed with him and friends the situation.

Sir William Beveridge promised to try and enlist the sympathies of one or two of the universities, and since his return to London he has been very active in this respect.

Although I cannot as yet say definitely what may or may not be the final result of the interviews which have taken place between Sir W. Beveridge and the Vice-Chancellors of London, Cambridge and Oxford universities, I feel certain that within a month or so we shall have an English group under the leadership of some outstanding personality who will undertake to raise funds, and I feel equally certain that such funds will be applied to good purpose.

I do not wish to interfere in any way with the formation of the English group which is entirely in the hands of English university people; nor can I represent such a group in any way. What I am concerned with at the present is to co-ordinate the foreign groups which are already in existence, and to stimulate the formation of groups in countries where there are no suitable groups as yet.

Of the different groups already in existence, I would like to mention the Committee of the Jewish Board of Deputies and Anglo-Jewish Association, appointed for the purpose of awarding fellowships to exiled Jewish scientists. Sir Philip Hartog is the Chairman of this Committee of which I have attended the first meeting.

Sir Philip Hartog will, I am convinced, see that nothing should interfere with the formation of a broader English group. I also had a long and satisfactory interview with Dr. Weizmann, to that effect.

All going well in England, I am free to leave for Belgium where I have an appointment with the Director of the Liege University, Mr. Duisberg,[17] on Saturday next,

---

[13]Frederick G. Donnan, professor at University College, London.
[14]Probably Baron Henry Mond Melchett, a director of Imperial Chemical Industries.
[15]Sir (Edward) John Russell, eminent government agricultural chemist.
[16]Possibly Max Delbrück, a physicist at the Kaiser-Wilhelm-Institut für Chemie, Berlin; but he tells us he does not recall seeing such a letter, although he knew of Szilard's efforts.
[17]J. Duisberg, rector.

May 13th in Brussels (I shall be at the Fondation Universitaire). I hope he will take up the matter with the other Belgian universities.

You probably know that Dr. Liebowitz[18] is in touch with the Anthropologist Franz Boas of Columbia University and that he has had an interview with Niels Bohr[19] in Copenhagen. He has arranged an interview between Bohr and Boas and hopes to hear soon from Boas about the steps which have been taken in the U.S.A. as a result of that interview.

I had conversations here with Niels Bohr, Harald Bohr,[20] Sir John Russell (Agricultural Chemistry), A. V. Hill (Physiology), Professor Hardy (Mathematics, Trinity College, Cambridge), and Donnan (Physical Chemistry). They all agree regarding the spirit in which constructive work should be carried on and would be glad to cooperate in one way or another if funds were available.

It seems now to be important that an international Board of some twenty scientists should be created, and I hope to have conversations wth some personalities on the subject whom we would wish to be chairman of such a Board. I hope I can do something along these lines before I leave for Belgium.

This is all the information I can give you today, and also I am not able to make any suggestions as to the details of how to co-operate with Bristol University. If you have an opportunity to discuss this matter with your friends there, please do so. I hope Dr. Liebowitz will be able to send you within the next 48 hours information of a confidential character. He will act instead of me during my absence from London in the next few days. He would also be glad if required to come to Bristol and discuss the matter with you personally, if the Vice-Chancellor should care to have more detailed information.

Yours very truly,
Leo Szilard

[Enclosure in preceding letter]

There are at present in various countries movements for raising funds to assist dismissed German scientists and scholars so as to enable them to continue their work as guests either as researchers or as lecturers at Universities.

It would be of importance to unite all these efforts with a view to the creation of an International framework which could be of permanent value. It has been suggested that an International Organisation should be set up which should be able to advise on the awarding of Fellowships to scholars and scientists who are at present without the means to continue their work. Fellowships would be granted if possible for a period of three to five years, and the scholars and scientists concerned would be allowed to work in any country of the world, so as to have a good chance of being permanently absorbed, in the course of time, in the countries to which they go.

---

[18] Benjamin Liebowitz, American physicist, inventor, and shirt manufacturer.
[19] Prominent Danish physicist.
[20] Niels' brother, a mathematician.

A possible way of co-ordinating the distribution of Fellowships would be to form immediately an International Board of some 20 to 50 scholars and scientists who would be willing to give advice if required on matters connected with the award of Fellowships, and to assume responsibility, each in his own field, for selecting those who should be awarded Fellowships. According to the Statutes of the suggested organisation, the Fellowships would be given to the most able men who are not in the position of pursuing their work, whatever their country of origin. At present, naturally, most of the Fellowships would be awarded to Germans who have to leave their country.

Such a Board is needed for the following reasons:

(a) The funds which will be raised in one country, for instance England, may be larger than needed for the small number of scientists who could conveniently work as guests in this country. There would be no point in crowding the laboratories in England with German scientists who could not be absorbed in the long run. Part of the funds raised in England should therefore be used for scientists and scholars who would do research work or who would lecture in less developed countries like India, Egypt, the Balkan States, etc.

(b) Some groups in some countries may emphasise in their effort to raise funds one or the other aspect of the acute situation, and thus become antagonistic to the German Government.

From the point of view of the scientists who may have a family living in Germany it is highly important that Fellowships *be not ear-marked* in any way. The scientist would be safe-guarded against this if he had to deal with a neutral International Board which could have among its members prominent German scientists who are in office at present.

Szilard comments on organizing the Academic Assistance Council, on the danger of accidental war, and on his personal situation:

Document 4

Imperial Hotel
Russell Square
London, W.C.1.
11th August, 1933

[addressee unknown]

I hope this letter will catch the mail. I have been working all day at the Academic Assistance Council (the English organisation which Sir William Beveridge, whom I met in Vienna, built up to place German scientists). They have appointed a young secretary[21] who is a very nice fellow and who will be efficient I hope, but who has gone

---

[21] Walter Adams.

away for four weeks, leaving the office in my care. Fortunately one of the lady secretaries[22] is excellent and I hope we shall manage to get useful things done in August. She is my invention in so far as I got her to come to London to this office from Geneva where I spent a few days and did some work in which she helped me. I was impressed by her ability and devotion and got the London office to take her on their staff. Now I get the benefit of my good deed, as I would be buried by the work without her being in the office. The real problems have not yet been attacked at all, and the office exhausted its energy in bureaucratic activity.

I am going to Cambridge tomorrow to arrange with Kapitza,[23] a Russian and a Fellow of the Royal Society, who is leaving for Russia next week, to take up the problem with the government there, and I hope that many of the scientists whom we cannot place in England will be able to work as experts in Russia, as there would be no point in crowding too many German scientists into England. Unfortunately I must be back in London on Sunday afternoon, so that my Cambridge visit will not be much of a rest.

In spite of being rather tired I feel very happy in England. This is partly due to the phenomenon that I always feel very happy for the first few months in a foreign country, but probably also due to the deeper sympathy I feel with the country and the people. I am not yet sure about the sympathy being mutual, but this is only a matter of practical importance.

The outlook in Europe is rather gloomy. It is quite probable that Germany will rearm and I do not believe that this will be stopped by intervention of other powers within the next years. Therefore it is likely to have in a few years two heavily armed antagonistic groups in Europe, and the consequence will be that we shall get war automatically, probably against the wish of either of the parties. Suppose if you have a large German and a large French air force, the false alarm is spread in Paris that the German air force has left the German airports, no French government, even the most pacifist one, could take the responsibility for holding back their air force to wait for confirmation of that rumour. The utmost they will do will be to make arrangement to call back their air force if the rumour subsequently turns out to be false. If they learn thereupon in Germany that the French aeroplanes have started, no hesitation whatever is conceivable in dispatching their air force in their turn.

I am afraid this is the most optimistic history of the next war and I will be astonished if it does not happen within the next five or ten years.

England and America are certainly the most hopeful two countries and they may or may not be out of the next war, but even if they keep out of it I do not think they can be proud of their aloofness.

I think most of my friends feel the burden of the situation and react by plunging deeper into their work and sealing hermetically their ears. I feel rather reluctant to follow their example, but I may have no choice left.

[22]Esther Simpson.
[23]Peter Kapitza, a prominent physicist who had been working at Cambridge.

By now practically all physicists who are any good have been placed and they have found out in Berlin [?] at last that I have done nothing for myself, so they tried to get a fellowship for me from an industrial group, but the resources of that group are exhausted for the time being. Of course it is impossible to apply for a fellowship for myself with those English committees on whose work I have a direct influence.

I am not against going to America, but I would very much prefer to live in England. I have not dismissed the idea of going to India, neither has this idea grown stronger. The fact that I am in close touch with Sir Philip Hartog should possibly be of some use if an opportunity in India arises.[24] I do not know if it would be wise to go to India for two years with a small English fellowship unless I were determined to stay there whatever happens.

I am spending much money at present for travelling about and earn of course nothing and cannot possibly go on with this for very long. At the moment, however, I can be so useful that I cannot afford to retire into private life.

It is almost ten o'clock at night and I have to stop.

You could send this letter on to Professor Bose,[25] Dacca University, whom your friends will certainly know. Give him my best regards and tell him both Bitter[26] (who happens to be in London) and I deplore his being absent.

Leo Szilard

The Japanese invasion of China spurred Szilard to attempt an abortive plan for a scientific boycott—an early example of his efforts to translate his concerns into action:

Document 5

Strand Palace Hotel
Strand
London, W.C.2.
24th April, 1934

Dear Lady Murray,[27]

When I last saw you we discussed the unfortunate situation which arose from the fact that no action was taken following the verdict of the League on the Japanese-Chinese conflict. As you will see from the enclosed page of to-day's 'Manchester Guardian' this subject is again arousing interest,[28] and I should be very glad if I could have your advice on a matter connected with this subject which is rather beyond my scope.

---

[24]Hartog was familiar with Dacca and other Indian universities.
[25]Probably S. N. Bose, professor of physics.
[26]Probably Francis Bitter, physicist at Westinghouse.
[27]Probably Mary Murray, daughter of the Earl of Carlisle and wife of (George) Gilbert Murray, president of the International Committee of Intellectual Cooperation and Chairman of the League of Nations Union.
[28]On the date of Szilard's letter the *Guardian* reported (p. 9) the Japanese rejection of any interference

Since the battle of Shanghai (I happened to be in New York at that time) [February, 1932] I had the opportunity of talking with several young scientists on the subject, and we prepared in New York the draft of a statement which I tried to reconstruct and which I am enclosing. We felt that a mere protest would not be of any value, but that a definite pledge on the part of the leading scientists, though of rather limited value in itself, would serve to "keep up faith" in the cause of justice.

The question which now arises is whether it would be advisable to get a well defined group (the Nobel Prize Winners) to sign such a pledge. It is possible that with two or three exceptions they would all be willing to sign it if none of them were asked to take any initiative in the matter.

If we came to the conclusion that such a pledge would be of value we would have to consider who would be the right person to ask the Nobel Prize Winners to lend their name to this cause. In asking them to do so one could point out from the beginning that the action will become void unless eight tenths of all the Nobel Prize Winners agree to sign the statement. Of course, whoever will take up the matter will have to rewrite the rather crude draft which I am attaching.

It would be of great help if I could have your comments on both these points, and I should appreciate if you could let me know on what days you will be in Oxford and whether you can see me; I should then soon go to Oxford for a day and telephone to you when I am there in order to fix a definite hour.

Yours sincerely,
Leo Szilard

[Draft of memorandum on the Sino-Japanese War]

The discoveries of scientists have given weapons to mankind which may destroy our present civilization if we do not succeed in avoiding further wars.

The temptation to resort to war would be less if the individual nations knew for certain that they would be compelled by international action to restore the status quo if they attained any of their objects by force. It is obvious that no status can be maintained forever and that sooner or later means will have to be found which will permit us to bring about peacefully those changes which will by then have become justified. However, as long as such means are not available only the rigid maintenance of the status quo against attempts at forcible changes can ensure security.

In Japan's conflict with China the scientist is in no position to know the rights or wrongs of either side. However, it is obvious that Japan has taken the law into its own hands, which no nation should be allowed to do. Other nations have been guilty of similar actions in the past, but this is no reason for tolerating such actions in the

---

by others in Chinese affairs: "Japan is now, with confident hope of success, asking the League [of Nations] and the Great Powers to capitulate again and give her a free hand to make China a Japanese Protectorate."

present or in the future. If in the case of the Japanese-Chinese conflict the status quo is not restored either by the good-will of Japan or by international action, Japan's example may soon be followed by others.

Therefore scientists should give careful consideration to the question as to whether they should join an international action of this kind by resorting to strict non-cooperation. This would involve the refusal to send scientific and technical information, including the periodicals, to the country against which the international action is directed and also the refusal to cooperate with students of that country.

One must hope that scientists in Japan will understand that there is no feeling against them and that they will undertake the difficult and ungrateful task of exerting their influence in favour of Japan's giving up any attempt at taking the law into its own hands.

The following group of documents covers Szilard's attempts to raise funds for experiments which might discover a nuclear chain reaction mechanism, and at the same time to create an organization of scientists that could exercise some influence over the consequences of such a discovery:

Document 6

6, Halliwick Road
London, N.10.
17th March, 1934

Dear Sir Hugo,[29]

As you are on holiday you might find pleasure in reading a few pages out of a book by H. G. Wells[30] which I am sending you. I am certain you will find the first three paragraphs of Chapter The First (The New Source of Energy, page 42) interesting and amusing, whereas the other parts of the book are rather boring. It is remarkable that Wells should have written those pages in 1914.

Of course, all this is moonshine, but I have reason to believe that in so far as the industrial applications of the present discoveries in physics are concerned, the forecast of the writers may prove to be more accurate than the forecast of the scientists. The physicists have conclusive arguments as to why we cannot create at present new sources of energy for industrial purposes; I am not so sure whether they do not miss the point.

It is perhaps possible to be more definite some time after your return, and in the meantime I hope you will in any case enjoy glancing through these few pages.

With best wishes for a pleasant stay,

Yours very truly,
Leo Szilard

---

[29] Sir Hugo Hirst, founder of the British General Electric Co., then at Carlton Hotel, Cannes.
[30] *The World Set Free.* "Chapter the First" begins with a prediction of the discovery in 1933 of means of "inducing radio-activity."

Document 7

[addressee unknown]   28th July, 1934

Memorandum of Possible Industrial Applications Arising out of a New Branch of Physics.

It is possible to indicate methods which might be successfully applied for the purpose of liberating atomic energy. It is not possible to foretell with certainty that these methods will be successful, but the experiments necessary for ascertaining this are fairly simple and could be carried out on a small scale in the university laboratories. Should such experiments give favourable results, the production of energy and its use for power production would be possible on such a large scale and probably with so little cost that a sort of industrial revolution could be expected; it appears doubtful, for instance, whether coal mining or oil production could survive after a couple of years.

I have applied for a group of patents in order to obtain patent protection for those methods which seemed to me promising, and it appears that these patents were successful in foreshadowing the latest developments in physics.[31]

They include, for instance, methods for the artificial production of radio-active bodies based on a process which recently has been discovered by Fermi.[32] The production of artificial "radium" for medical purposes based on these processes seems to be a sound commercial proposition, but it would be sidetracking the issue to concentrate on this point.

Facilities are required for two different purposes:—

1. In order to develop and maintain a group of valid patents £500 are required for the next year, which would also take care of administrative expenses connected with the maintenance of the patents.

2. If we wish to start the necessary experiments one ought to secure the continuity of work for two to three years. It is not possible to state exactly what facilities will be required as this will depend to a large extent on what facilities will be provided by the university laboratory which would be used as a frame for this work. It would, however, be advisable to have £2,000 available for expenditure that may be incurred.

From the point of view of a financier who could consider contributing to the required facilities the position is this:—the chances that the envisaged experiments will yield a favourable result may be estimated to anything between 1 to 20 and 1 to 5. The value of the return in case of success is, of course, enormous and could hardly be estimated in terms of money, so that from the purely financial point of view it is a sort of lottery with a fairly good chance to win a prize and enormous prizes. Yet it would

---

[31] See *Scientific Papers*, pp. 605–651.
[32] The process is bombardment with neutrons.

be highly preferable to get financial support from quarters that would consider the experiment as a research work in the field of science which has a good chance of highly significant industrial applications, and realise that the exploitation of discoveries of this scope must not be organised on a purely commercial basis.

Difficulty will undoubtedly arise from the fact that it is not easy for anybody to form an independent opinion of his own on the merits of the case. A possible way out would be to get the opinion of some of the professors of the University of London who are working themselves in this field, and with whom I can easily keep in touch on the matter.

Document 8

ABSCHRIFT [copy]

Von [from]: M. Polany[i]
Kenmore
Didsbury Park
Manchester
Tel. 3838 Didsbury
11 Nov. 1934

Dear Szilard,

I have spoken to Donnan about you when he was staying with us during the past two days. I have mentioned no details, it was he who did the speaking mostly. He would like to help you. His idea is a financier, a very rich man, who would let you do just as you please, asking only for a dividend at the end. Do you want him to follow up this direction? Personally I do not feel too confident of the success, but the plan is not unintelligent. Furthermore: will you let me report to Aschner that you are in the course of making great inventions and that I would see a favourable opportunity for an investment as *stiller Gesellschafter* [silent partner]. He could ask me and Donnan to be his trustees. Especially if you come to work in Manchester. These overcautious people are often the first to invest their money under such extraordinary conditions.

Donnan told me that there is an opposition to you on account of taking patents. A physicist (not Rutherford) told him so.

Please answer question about Aschner soon, since I am about to write him about something else.

Yours,
M. P.

Szilard redoubled his efforts when he began (mistakenly) to believe that indium might have the chain reaction mechanism he sought:

Document 9

PERSONAL

74, Gower Street
London, W.C.1.
3rd June, 1935

Professor Lindemann
Clarendon Laboratory
Oxford

Dear Professor Lindemann,

I hope very much to see you on Wednesday and talk to you about a matter which appears to me to be of great seriousness. For some time back I have suspected that the three radio-active periods which Chalmers and I found in the case of indium involved a new type of process in which a neutron

a) either knocks out another neutron from indium 113 in a non-capture process, or

b) liberates a neutron of the mass number 2 from indium 113 and gets captured in the process.

I have gradually come to the conviction that either a process of the type a) or, alternatively, a process of the type b) does occur and is possibly responsible for a number of other known radio-active periods, which I believe I can single out. I believe you will share this conviction after you have heard my arguments on Wednesday.

The question whether a neutron of the mass number 2 exists and can be liberated by fast neutrons cannot be answered offhand, but it is perhaps fair to say that since one of the two processes a) or b) certainly occurs, we have something like a fifty to fifty chance that such "double neutrons" are involved.[33]

It seems to me that the question whether or not the liberation of nuclear energy and the production of radio-active material on a large scale can be achieved in the immediate future, hinges on the question whether or not "double neutrons" can be produced. If "double neutrons" can be produced, then it is certainly less bold to expect this achievement in the immediate future than to believe the opposite.

Even if I am grossly exaggerating the chances that these processes will work out as I envisage it at present, there is still enough left to be deeply concerned about what will happen if certain features of the matter become universally known. In the circumstances, I believe an attempt, whatever small chance of success it may have, ought to be made to control this development as long as possible.

There are two ways in which this can be attempted. The more important one is secrecy, if necessary, attained by agreement among all those concerned that another form of publication should be used as far as the dangerous zone is concerned, which would make experimental results available to all those who work in the nuclear field

---

[33]In fact there is no neutron isotope of mass 2; Szilard was led astray by the fact that indium has isomers, i.e., radioactive excited states of the same mass as the normal state.

in England, America and perhaps in one or two other countries, but otherwise keeping the result quiet, until those who are concerned are satisfied that no "double neutron" is involved.

The other way, the less important one, is to take out patents. Early in March last year it seemed advisable to envisage the possibility that, contrary to current popular opinion, the release of large amounts of energy and the production of large amounts of radio-active material might be imminent. Realising to what extent this hinges on the "double neutron", I have applied for a patent along these lines, including also the production of radio-active material by neutron bombardment. This was filed before Fermi started his fundamental experiments and was followed by a number of further patent applications along the same lines. Obviously it would be misplaced to consider patents in this field private property and pursue them with a view to commercial exploitation for private purposes. When the time is ripe some suitable body will have to be created to ensure their proper use. Also one has to avoid applying for patents wherever secrecy is endangered or in countries which are likely to misuse them; so far I have carefully observed this point.

Though I do not know for the present what will be the proper steps in this matter, I am very anxious to keep my full freedom of action in everything connected with it.

As far as experiments in this special field go, I should like to keep them, as far as possible, in my own hands and not merely act as a "catalyst". As long as Collie, Griffiths[34] and I work alone in this field at Oxford, it is not quite easy for me to run these experiments in my own way and, without appearing pretentious, publish or not publish, according to what I think I should. I hesitate also to suggest that the whole Nuclear Department at Oxford should work in a field which may yield very little of purely scientific interest, if it turns out that we have to deal with a non-capture process after all.

If I knew that it would be convenient to you, I should make an attempt to get a budget of £1,000 for next year from private persons in order to be able to take on one or two helpers with whom I could work in this special field in the Clarendon Laboratory, while I would still, if it appears useful, work with Collie and Griffiths in the general field as envisaged hitherto.

Whether an attempt to get financial assistance will be successful or not, I cannot tell, but I feel justified in approaching men of vision in this matter. I should be very happy if you, too, thought that Oxford is in many ways well suited for this type of work and that, conversely, this type of work could greatly accelerate the building up of nuclear physics in Oxford. . . . [35]

Yours sincerely,
Leo Szilard

---

[34]C. H. Collie and J. H. E. Griffiths, researchers at the Clarendon.
[35]A final paragraph dealing with publication matters is omitted.

Document 10

Professor M. Polanyi
The Victoria
University of Manchester
Department of Chemistry
Telephone: Ardwick 2681
28th June, 1935.

My dear Szilard,

I gather from my talk with Weizmann that he is favourably inclined towards the foundation of a research corporation on the lines discussed by us in Manchester. He will raise this point with Willstätter[36] by the end of this week, and will probably ask me to come to London for a discussion on this matter. I have promised to go if he requires me; so we might meet soon again.

Yours,
M. P.

Dr. L. Szilard
The Clarendon Laboratory
Oxford

Document 11

[Ch. Weizmann]
16, Addison Crescent
London, W.14.
5th July, 1935

Professor M. Polanyi
Manchester University
Manchester

My dear Professor Polanyi,

Willstätter was here for only a very short time, and I felt it would have been too much trouble to ask you to come to London. But I had a word with him with regard to Dr. Szilard's affairs, and I think he shares your view. I shall take some steps in the near future, and shall let you know if anything develops as a result of them.

With kind regards,
I am,
Very sincerely yours,
Ch. Weizmann

---

[36]Possibly Richard Willstätter, prominent German chemist working in Zürich.

Document 12

[M. Goldhaber][37]
46, Belvoir Rd.
Cambridge
18.3.36

Dear S.,

Somebody sent a copy of your neutron-patents to the Lab. It was freely discussed at tea and, of course, your intentions were misunderstood to be financial or otherwise unscientific. I defended you by saying that in your opinion it is better to direct technical applications of science in a way which will be useful to science and not damaging to the public. There is no reason for you to worry about this.

In great hurry

Yours,
M. G.

Szilard noticed something strange about uranium:

Document 13

c/o Clarendon Laboratory
Parks Road
Oxford
26th March, 1936

Dear Professor Bohr,

Might I add to our last conversation that I have talked to Kalckar[38] and Placzek[39] before they left London about the same things about which I talked to you, so should you happen to wish to discuss any points which may occur to you in connection with this matter with either of them, you need not have any hesitation about it.

I wish to draw your attention to the paper of Hahn and Meitner in the last "Naturwissenschaften".[40] They overlook the fact that there exists a stable isotope of uranium (mass number 235) which has been discovered by Dempster in the course of this year.[41] The case of uranium seems to me somewhat analogous to the case of indium, though certainly not similar. I find it difficult to assume so many isotopic isobars as we would have to assume in the case of uranium if we wanted to put all the blame on isotopic isobars.[42]

[37]Maurice Goldhaber, physicist, at that time at Cambridge, England.
[38]Possibly F. Kalckar, nuclear physicist, who moved to California about this time.
[39]George Placzek, emigré Czech physicist.
[40]Vol. *24*: 158–159 (March 6, 1936). This paper gives a table of ten supposed isotopes, from Th235 up to "Eka-Ir239", all resulting from neutron bombardment of uranium.
[41]Arthur J. Dempster, professor of physics at the University of Chicago.
[42]i.e., on isomers. Hahn and Meitner were indeed mistaken; they were observing the products of nuclear fission and calling them isomers of uranium and other heavy elements.

Should you come to any conclusion on the subject of isotopic isobars and be able to give an upper limit for the half life period, I should appreciate your letting me know about it.

With best wishes,

Yours sincerely,
Leo Szilard

The following pleas to two leading Cambridge physicists show Szilard's ideas, already well developed in 1936, on measures for controlling the exploitation of nuclear chain reactions. Because of reservations towards his scientific and social ideas already expressed at Cambridge, he explained himself carefully.

Document 14

c/o Clarendon Laboratory
Parks Road
Oxford
27th May, 1936

Dear Lord Rutherford,

Enclosed you will find a draft for a letter to "Nature" which I am sending to you in the hope that it might interest you.[43] Although it was written some time ago, I still hesitate to send it in for publication and I should like to mention here the reasons for this hesitation:

I have so far not been able to exclude the possibility that a heavy isotope of the neutron of mass number 4 exists and is involved in the anomalous Fermi effect of indium. It can easily be shown that such a particle would have a mass larger than 4.014. In the circumstances—if a heavy neutron isotope really exists—we would for the first time have to envisage the theoretical possibility of nuclear chain reactions. The prospect of bringing about nuclear transmutations on a large scale by means of such chain reactions is somewhat disconcerting. It is very unlikely that the misuse of chain reactions could be prevented if they could be brought about and became widely known in the next few years. I am quite aware that the view which I am taking on the subject may be very exaggerated. Nevertheless, the feeling that I must not publish anything which might spread information of this kind—however limited—indiscriminately, has so far prevented me from publishing anything on this subject. Since I am not quite clear about the proper course to take, I thought of discussing the matter with Dr. Cockroft,[44] to whom I am going to write to-day.

If I were convinced that the matter is really important, I should perhaps not hesitate to ask for your advice at this juncture. Being at the moment unable to estimate the real importance of the matter, I shall ask Dr. Cockroft, if I see him, to inform you

---

[43]This went through several versions. Szilard submitted a letter to *Nature*, "Anomalies in Radioactivity Induced by Neutrons," on November 5, 1936, but this was later withdrawn and never published.
[44]John D. Cockroft, physicist at Cambridge.

about our conversation if he thinks all this to be sufficiently important. I shall of course be glad to come to Cambridge at any time, in case you are disposed to express an opinion or desire to have further information on this subject. All this applies equally to a second matter which I equally hope to discuss with Dr. Cockroft:

You might perhaps remember that I mentioned to you in passing some two years ago patents for which I had applied in March 1934. One such patent has now been granted on the principle of producing artificial radio-active elements by neutrons and the question arises, to what use, if any, such patents ought to be put.[45]

I cannot consider patents relating to nuclear physics as my property in any sense whatever. It would seem that if such patents are important, they ought to be administered in a disinterested way by disinterested persons. It is, however, hardly for me to take any irreversible steps about these patents which I have taken out in the capacity of a sort of self-appointed trustee. Apart from yourself, one could perhaps think of Chadwick,[46] Cockroft, Joliot and Fermi as being the persons the most entitled to say whether these patents should be withdrawn or maintained and in what way and by whom they should be administered, if they are maintained.

Soon after Fermi's first discovery I discussed with Dr. Oliphant my reasons for thinking that the existence of such patents might be useful. I am referring to this in greater detail in the enclosed copy of a letter addressed to Dr. Cockroft, which I am sending you in case you should care to have more detailed information.

Yours very truly,
Leo Szilard

Document 15

c/o Clarendon Laboratory
Parks Road
Oxford
27th May, 1936

Dear Dr. Cockcroft,

Please forgive me for being unable to talk about slow neutrons or any other similar subject in public (Kapitza Club) this term. There are, however, some unpublished observations which I made in January and which I should very much like to discuss with you. They may or may not have a direct bearing on another matter on which I should like to have your advice. This concerns patents for the production of artificial radio-active elements for which I applied during a period of enforced leisure between March and September 1934.

I am enclosing a copy of a letter which I have sent to Lord Rutherford for your information and I should like to add the following:—

---

[45]British Patent 440,023. See *Scientific Papers*, pp. 622-638.
[46]James Chadwick, physicist at Cambridge.

It would not seem right to me that physicists who take out such patents should derive financial or other privileges from them and some form of disinterested control ought to be found, if industrial applications are considered to be of some importance. I am enclosing a booklet on the American Research Corporation[47] which might interest you in this connection, although I do not believe that their example ought to be too closely imitated.

When I applied for the first patent in March 1934, I somewhat foolishly thought such patents might simply be asigned to the Cavendish.[48] After Fermi's first discovery, I went to see Dr. Oliphant and explained to him my reasons for thinking that the existence of such patents might be useful, if they are controlled by the proper persons. (Subsequently the patent relating to Fermi's discovery has, along with others, been offered to a Government Department; although it was made clear that the question of a financial compensation does not arise in any way, this particular patent has not been accepted and has been published under my name.) I also told Dr. Oliphant that in my personal opinion the possibility of an important industrial development depends on the possibility of setting up enormously efficient sources of neutrons and that this possibility in its turn depends on the question of the existence of a heavy isotope of the neutron.

If such multiple neutrons exist, we may envisage, if we wish to do so, the theoretical possibility of an industrial revolution in a not too distant future. In that case patents might be used by scientists in a disinterested attempt to exercise some measure of influence over a politically dangerous development.

On the other hand, if no heavy neutron isotope exists, it would seem that an industrial development based on the application of nuclear physics must necessarily be very limited and there is no real need for any physicist to concern himself with such patents. The only use to which such patents could then be put, is towards obtaining funds for research purposes from persons interested in the promotion of industrial development. I personally felt inclined to think that good use could be made of such funds if they were forthcoming.

I should very much like to have your opinion whether you think that some form of disinterested control of such patents can be found, since if this does not appear to be feasible, I personally should like to withdraw those patents which I have taken out.

In spite of the uncertainty of the assumption that a heavy neutron isotope is involved in my experiments, I have no choice but to try to investigate the matter by more direct methods of observation. I am not certain that the Wilson Cloud Chamber is the most suitable instrument in this case and I should very much like to have Dee's[49] opinion on this question. Would you be kind enough to pass this letter with the enclosure on to him? I may be in Cambridge on Saturday and Monday morning. Could

---

[47]The Research Corporation, founded by James Cottrell; it supported research with money derived from patents.
[48]Cavendish Laboratory, Cambridge.
[49]P. I. Dee, expert on cloud chambers at Cambridge.

you let me know if you will be about on one of these days? I should then also attempt to get hold of Dee, if he happens to be free.

With best wishes,

Yours sincerely,
Leo Szilard

In 1938 the little money Szilard had earned from consulting, patent arrangements, and the like was running out, and at the same time Imperial Chemical Industries Ltd., which had supported him with a fellowship, decided to halt its support of refugee scientists. The following documents indicate his concern over both his personal situation and the political situation in Europe.

Document 16

WESTERN UNION (night letter) [October, 1938]

LINDEMANN
CHRIST CHURCH
OXFORD, ENGLAND

HAVE ON ACCOUNT OF INTERNATIONAL SITUATION WITH GREAT REGRET POSTPONED MY SAILING FOR AN INDEFINITE PERIOD STOP WOULD BE VERY GRATEFUL IF YOU COULD CONSIDER ABSENCE AS LEAVE WITHOUT PAY STOP WRITING STOP PLEASE COMMUNICATE MY SINCERELY FELT GOOD WISHES TO ALL IN THESE DAYS OF GRAVE DECISIONS

SZILARD

Document 17

c/o Liebowitz
420 Riverside Drive
New York City
October 21st, 1938

Dear Tuck[50]

I wonder whether you see the European situation in the same way as I do. I believe that this last breach of faith which led to the Munich agreement has definitely settled the fate of Europe for a long time to come. It will not be possible for either England or France to make international agreements and get anybody to believe that they will keep these agreements, if keeping them means risking war. It follows that international anarchy will rule in Europe and perhaps elsewhere. This may mean war or merely continuous unrest accompanied by "peaceful" changes.

I do not think that in the circumstances I would wish to live in England if I am limited to do work in physics in the Clarendon. But I would return to England if I

---

[50]James L. Tuck, physicist at Oxford.

find that it will be possible for me to cooperate with others in working towards changing this disastrous situation. I am taking steps to find out if anything of such nature is possible, in the meantime I shall ask Professor Lindemann for a further leave of absence without pay, or, if that should prove impossible, I shall resign from the Clarendon.

This for your personal information only. As to yourself, knowing that you are not very much interested in the fate of nations and believe that nothing can be done about it anyway, I think that you are very wise to think of America. Maybe you could try for a Common Wealth Fellowship for next year. Whether I can find something for you here right away I do not know, but I would try it at once and with great energy if, having received this letter, you advise me by cable that you want me to do so. I feel that it is doubtful whether you should accept any position which is not safe for at least two years and carries a salary of at least $2500. I might try to find something for you on such terms on the basis of our cystein work,[51] as it seems that I can make people in this country enthusiastic about it. Of course, I do not know in the least whether I can get them to go at it seriously. I believe though that another year or two spent with radioactive indicator work with cystein, carried out in collaboration with physiological chemists, would give you a very good start here. There is some danger though that, if I talk to people about you, mentioning your name—and of course I would have to mention it—rumours might arise which could reach Oxford and perhaps prejudice your position. You must let me know whether you want to take that risk. I shall then take all precautions I can. As to the merits of the question whether you should or should not leave England, I have little to say. Unless you take a Common Wealth Fellowship and come here as a temporary visitor, leaving England means a very grave decision. If you decide at all that you want to leave England, you ought to take that decision either on the ground that you believe that you would rather live here for the next ten years than in England, or on the ground of some broad principle. Moreover, it can easily be that, in spite of your decision, I am not able to find anything suitable for you. Of course, the possibility to apply for a Common Wealth Fellowship would still remain. If you have something definite to say about wanting me to do something, please send me a cable, because I have some free time now and intend anyway to discuss cystein with a number of people in the near future. In a few weeks time I might be away from the East and unable to do much about it.

As to the cystein publication I am not certain whether it is wise to say something about "moonshine", but if you prefer a joint publication I shall reconsider the matter. There is one condition, however, on which I have to insist, that is that I may make a footnote pointing out that you carried out all observations yourself during my absence from Oxford. This might be unconventional, but I feel that I cannot take credit for what I have not done. If you think we can make such a footnote, then I would be

---

[51]See L. Tuck, "Radioactive Sulphur for Biochemical Experiments," *J. Chemical Society London, 1939*:1292–1294 (1939), where Tuck thanks Szilard "who foresaw a use for this work." Cystein is an amino acid.

very glad to publish jointly but would have still to think a little bit about the advisability of a joint publication from the point of view that it might be premature to say those things which I wanted to say. I quite agree that you have to publish in any case very soon in order to satisfy the Salter people. By the way, what is your status now? Have they not extended the fellowship for another year? Or did Lindemann make some other arrangement? What do you think are your chances of a real career at Oxford?

As to the electron transformer, the instability of the electron path in a radially decreasing field of cylindrical symmetry which you think you have proved, does not exist. The mistake you make is the following: . . .[52]

Though there is no instability of the electron path, there is an instability of the political situation, and if you also feel that we should not build an electron transformer at Oxford, please let me know quickly so that I can officially tell Lindemann that we are abandoning the project.

Please write.

Yours,
Leo Szilard

Document 18

c/o Liebowitz
420 Riverside Drive
New York City
January 13th, 1939

Professor F. A. Lindemann
Christ Church
Oxford, England

Dear Professor Lindemann:

Three months have now passed since, acting on an impulse, I cabled you that I am postponing my sailing for an indefinite period on account of the international situation, and that I should be grateful if my further absence could be considered as leave without pay.[53] I sent you this cable after Czechoslovakia was forced to accept the Berchtesgaden demands [early October, 1938], and it must have reached you at a moment when many people believed that there was an immediate danger of general war. You may have therefore thought that this assumed danger prompted me to postpone my sailing, and you may then have wondered why I did not return to Oxford after the Munich agreement, that is if you gave any thought to my continued absence at a time when urgent political and defense questions must have been claiming most of your thoughts.

---

[52]We omit the rest of the paragraph, which discusses motion of electrons in a cyclotron or similar device.
[53]See *Doc. 16*.

It seems to me that the Munich agreement created, or at the very least demonstrated, a state of international relations which now threatens Europe and in the long run will threaten the whole civilized world. This cannot fail to claim the attention of all of us, and, if the situation is to be improved, the active cooperation of many of us. I greatly envy those of my colleagues at Oxford who in these circumstances are able to give their full attention to the work which has been carried on at the Clarendon Laboratory and who are able to do so without offending their sense of proportions. To my great sorrow I am apparently quite incapable of following their example.

Since my collaboration in the work, for which you were good enough to win the support of Imperial Chemical Industries, would be of little value unless I gave the work my full attention, it seems best in the circumstances that I should not embark upon it. This being so, I do not feel that I am entitled to keep any payments which Imperial Chemical Industries may have made to me under the new agreement, i.e. after January 1st of last year. I should be grateful if you could perhaps communicate on this subject with Dr. Slade[54] and tell him how very thankful I am for the help I had from Imperial Chemical Industries in the past, and how very much I regret that the deterioration of the international situation, which occurred while I was abroad, makes it impossible for me to collaborate in the work which Dr. Slade kindly consented to support. If Dr. Slade wishes me to refund payments made to me after January 1st of last year, I shall be very glad to do so. In this case Dr. Slade will have to let me know the amount which actually has been paid to my account, and also to what account and under what heading he wishes me to transfer this amount.

It seems to me that those who wish to continue to dedicate their work to the advancement of science would be well advised to move to America where they may hope for another ten or 15 years of undisturbed work. I myself find it very difficult, though, to elect such "individual salvation", and I may therefore return to England if I can see my way of being of use, not only in science, but also in connection with the general situation. It is hardly necessary to state that, if I shall be in England and if you want me to do so, I shall be most happy again to cooperate with those who work in the Clarendon Laboratory. It may be best, however, that I should not receive financial support from the Laboratory, as such financial support is bound to be linked with fixed obligations which I would rather avoid.

For the time being, I do not yet see my way of being of use in England in connection with the general situation, though I see certain potential possibilities in this respect. In view of these I am at present not looking for a "job" on this side of the Atlantic. Perhaps I shall have an opportunity to talk to you about all this if I shall visit England in a not too distant future.

Naturally I regret that it will not be possible for me to collaborate in building up apparatus for the new Clarendon Laboratory. I trust that the spirit of inflation, which must necessarily accompany any armament race such as is on at present, will at least

[54]Dr. R. E. Slade, engineer and administrator at I.C.I.

make it possible for you to obtain the funds necessary for carrying on research in the new laboratory.

Please excuse the three months' delay of this letter. Immediately after the Munich agreement it did not seem possible for me to have a sufficiently balanced view, and I had to allow some time to elapse before I was able to write without bitterness of this event.

With kind regards to all, I am,

Yours very sincerely,
Leo Szilard

Chapter II
There was very little doubt in my mind that the world was headed for grief.

I was still intrigued with the possibility of a chain reaction, and for that reason I was interested in elements which became radioactive when they were bombarded by neutrons and where there were more radioactive isotopes than there should have been. In particular I was interested in indium. I went up to Rochester [New York] and stayed there for two weeks and made some experiments on indium which finally cleared up this mystery. It turned out that indium is not unstable and that the phenomenon observed could be explained without assuming that indium is split by neutrons.[1]

At that point I abandoned the idea of a chain reaction and I abandoned the idea of looking for elements which could sustain a chain reaction. I wrote a letter to the British Admiralty suggesting that the patent which had been applied for should be withdrawn because I couldn't make the process work. Before that letter reached them, I learned of the discovery of fission. (*Docs. 19–21*) This was early in January when I visited Mr. [Eugene] Wigner in Princeton, who was ill with jaundice.

Wigner told me of [Otto] Hahn's discovery. Hahn found that uranium breaks into two parts when it absorbs a neutron; this is the process which we call fission. When I heard this I saw immediately that these fragments, being heavier than corresponds to their charge, must emit neutrons, and if enough neutrons are emitted in this fission process, then it should be, of course, possible to sustain a chain reaction. All the things which H. G. Wells predicted appeared suddenly real to me.

At that time it was already clear, not only to me but to many other people—certainly it was clear to Wigner—that we were at the threshold of another world war. And so it seemed to us urgent to set up experiments which would show whether in fact neutrons are emitted in the fission process of uranium. I thought that if neutrons are in fact emitted in fission, this fact should be kept secret from the Germans. So I was very eager to contact Joliot[2] and to contact Fermi,[3] the two men who were most likely to think of this possibility.

I was still in Princeton and staying at Wigner's apartment—Wigner was in the hospital with jaundice. I got up in the morning and wanted to go out. It was raining cats and dogs. I said, "My God, I am going to catch a cold!" Because at that time, the first years I was in America, each time I got wet I invariably caught a bad cold. However, I had no rubbers with me, so I had no choice, I just had to go out. I got wet and came home with a high fever, so I was not able to contact Fermi. As I got ready to go

---

[1] M. Goldhaber, R. D. Hill, and L. Szilard, *Physical Review*, 55:47–49 (1939) in *Scientific Papers*, pp. 155–157.
[2] Frédéric Joliot, professor of nuclear chemistry at the Collège de France.
[3] Enrico Fermi, who had immigrated from Italy and taken a professorship of physics at Columbia University a few weeks before.

back to New York, I opened the drawer to take my things out and saw there were Wigner's rubbers standing. I could have taken Wigner's rubbers and avoided the cold. But as it was I was laid up with fever for about a week or ten days. In the meantime, Fermi also thought of the possibility of a neutron emission and the possibility of a chain reaction and he went to a private meeting in Washington and talked about these things. Since it was a private meeting, the cat was not entirely out of the bag, but its tail was sticking out.

When I recovered I went to see Rabi,[4] and Rabi told me that Fermi had similar ideas and that he had talked about them in Washington. Fermi was not in, so I told Rabi to please talk to Fermi and say that these things ought to be kept secret because it's very likely that neutrons are emitted, this may lead to a chain reaction, and this may lead to the construction of bombs. So Rabi said he would and I went back home to bed in the King's Crown Hotel. A few days later I got up and again went to see Rabi, and I said to him, "Did you talk to Fermi?" Rabi said, "Yes, I did." I said, "What did Fermi say?" Rabi said, "Fermi said 'Nuts!'" So I said, "Why did he say 'Nuts!'?" and Rabi said, "Well, I don't know, but he is in and we can ask him." So we went over to Fermi's office, and Rabi said to Fermi, "Look, Fermi, I told you what Szilard thought and you said 'Nuts!' and Szilard wants to know why you said 'Nuts!'" So Fermi said, "Well . . . there is the remote possibility that neutrons may be emitted in the fission of uranium and then of course perhaps a chain reaction can be made." Rabi said, "What do you mean by 'remote possibility'?" and Fermi said, "Well, ten per cent." Rabi said, "Ten per cent is not a remote possibility if it means that we may die of it. If I have pneumonia and the doctor tells me that there is a remote possibility that I might die, and it's ten per cent, I get excited about it."

From the very beginning the line was drawn; the difference between Fermi's position throughout this and mine was marked on the first day we talked about it. We both wanted to be conservative, but Fermi thought that the conservative thing was to play down the possibility that this may happen, and I thought the conservative thing was to assume that it would happen and take all the necessary precautions. I then went and wrote a letter to Joliot, in which I told Joliot that we were discussing here the possibility of neutron emission of uranium in the fission process and the possibility of a chain reaction, and that I personally felt that these things should be discussed privately among the physicists in England, France, and America, and that there should be no publication on this topic if it should turn out that neutrons are, in fact, emitted and that a chain reaction might be possible. This letter was dated February 2, 1939. (*Doc. 29*) I then sent a telegram to England asking them to send a block of beryllium which I had had made in Germany with the kind of experiments in mind which I now was actually going to perform.

Such a block of beryllium can be used to produce slow neutrons, because if you put radium in the middle of it, under the influence of the gamma rays of radium the beryllium splits and gives off slow neutrons. If uranium in the process of fission, which can

---

[4]Isidor Isaac Rabi, professor of physics, Columbia University.

be caused by slow neutrons, emits fast neutrons, these fast neutrons can be distinguished from the neutrons of the source by virtue of their higher energy.

There was at Columbia University some equipment which was very suitable for these experiments. This equipment was built by Dr. Walter Zinn, who was doing experiments with it. All we needed to do was to get a gram of radium, get a block of beryllium, expose a piece of uranium to the neutrons which come from beryllium, and then see by means of the ionization chamber which Zinn had built whether fast neutrons were emitted in the process. Such an experiment need not take more than an hour or two to perform, once the equipment has been built and if you have the neutron source. But of course we had no radium.

So, I first tried to talk to some of my wealthy friends, but they wanted to know just how sure I was that this would work. Finally I talked to one of my not-so-wealthy friends. He was a successful inventor, he was not poor but he was not exactly wealthy. His name is Benjamin Liebowitz. He derived some income from royalties. I told him what this was all about, and he said, "How much money do you need?" I said, "Well, I'd like to borrow $2,000." He took out his checkbook, he wrote out a check, I cashed the check, I rented the gram of radium, and in the meantime the beryllium block arrived from England. *(Docs. 23–25)* With this radium and beryllium I turned up at Columbia and, having talked to Zinn, said to the head of the department, "I would like to have permission to do some experiments." I was given permission to do experiments for three months. I don't know what caused this caution, because they knew me quite well; but perhaps the idea was a little too fantastic to be entirely respectable. And once we had the radium and the beryllium it took us just one afternoon to see the neutrons. Mr. Zinn and I performed this experiment. [2]

[On March 3, 1939] everything was ready and all we had to do was to turn a switch, lean back, and watch the screen of a television tube. If flashes of light appeared on the screen, that would mean that neutrons were emitted in the fission process of uranium and this in turn would mean that the large-scale liberation of atomic energy was just around the corner. We turned the switch and we saw the flashes. We watched them for a little while and then we switched everything off and went home. That night there was very little doubt in my mind that the world was headed for grief. [8]

At that point I thought that from this point on there should be no difficulty in obtaining financial support for this work. But in this I was quite mistaken. [1]

In the meantime Fermi, who had independently thought of this possibility, had set up an experiment. His did not at first work so well, because he used a neutron source which emitted fast neutrons, but then he borrowed our neutron source and his experiment, which was of a completely different design, also showed the neutrons.[5]

Soon after we had discovered the neutron emission of uranium, Wigner came to New York and we met—Fermi and I and Wigner—in the office of Dr. Pegram.[6]

---

[5]H. L. Anderson, E. Fermi, and H. B. Hanstein, *Physical Review,* 55:797–798 (1939).
[6]George B. Pegram, chairman of the Physics Department and dean of the Graduate Faculties, Columbia University.

Wigner said that this was such a serious business that we could not assume the responsibility for handling it; we must contact and inform the government. Pegram said that we could do that; he knew Charles Edison, who was the assistant secretary of the navy.[7] He told Edison that Fermi would be in Washington the next day and would be glad to meet with a committee and explain certain matters which might be of interest to the navy. Fermi went there. He was received by a committee. He told in his cautious way the story of uranium and what potential possibilities were involved. But there the matter ended. I got an echo of this through Merle Tuve.[8] Ross Gunn,[9] who was an adviser to the navy and who attended this conference, telephoned Tuve and asked him, "Who is this man Fermi? What kind of a man is he? Is he a Fascist or what? What is he?" Nothing came of this.[10]

And now there came the question: Shall we publish this? There were intensive discussions about this. Zinn and I, and Fermi and Anderson,[11] each sent a paper to the *Physical Review*, a letter to the editor.[12] But we requested that these publications be delayed for a little while until we could decide whether we wanted to keep these things secret or whether we would permit them to be published. Throughout this time I kept in close touch with Wigner and with Edward Teller,[13] who was in Washington. I went down to Washington; Fermi also went down to Washington on some other business, I forget what it was, and Teller and Fermi and I got together to discuss whether or not these things should be published.[14] Both Teller and I thought that they should not. Fermi thought that they should. But after a long discussion, Fermi took the position that after all this was a democracy; if the majority was against publication, he would abide by the wish of the majority, and he said that he would go back to New York and advise the head of the department, Dean Pegram, to ask that publication of these papers be indefinitely delayed.

While we were still in Washington we learned that Joliot and his coworkers had sent a note to *Nature*, reporting the discovery that neutrons are emitted in the fission of uranium and indicating that this might lead to a chain reaction.[15] At this point

---

[7] This decision was made March 16, 1939, immediately following Germany's annexation of Czechoslovakia. Pegram did not speak directly to Edison but made arrangements by telephone with Edison's office, according to [8], p. 3.
[8] Merle A. Tuve, physicist at the Carnegie Institution of Washington's Department of Terrestrial Magnetism.
[9] Technical advisor to the Naval Research Laboratory.
[10] In fact Gunn began work on separation of uranium-235 as a result of this conference. In the original tapes this paragraph came later, out of chronological order; see n. 12 in chapter III, Recollections. For the original order see Szilard's "Reminiscences" in *Perspectives in American History*, 2:94–151 (1968).
[11] Herbert Anderson, graduate student working with Fermi.
[12] Szilard and Walter H. Zinn, *Physical Review*, 55:799–800 (1939), in *Scientific Papers*, pp. 158–159. H. L. Anderson, E. Fermi, and H. B. Hanstein, *Physical Review*, 55:797–798.
[13] Physicist at George Washington University; like Szilard and Wigner, a Hungarian-born immigrant.
[14] March 19, 1939. Fermi's business included the meeting with naval officers, March 17, described above.
[15] Hans von Halban, Jr., F. Joliot, and Lew Kowarski, *Nature*, 143: 470–472 (1939). Szilard heard of this ca. March 20, according to [8]. For the reactions of the French and other details of the affair see Spencer Weart, "Scientists with a Secret," *Physics Today*, 29: 23–30 (February, 1976); Weart, *Scientists in Power. France and the Origins of Nuclear Energy, 1900–1950* (Cambridge, Mass.: Harvard University Press, in preparation), chapter 6.

Fermi said that in this case we were going to publish now everything. I was not willing to do that, and I said that even though Joliot had published this, this was just a first step, and that if we persisted in not publishing, Joliot would have to come around; otherwise he would be at a disadvantage, because we would know his results and he would not know of our results. However, from that moment on Fermi was adamant that withholding publication made no sense. I still did not want to yield and so we agreed that we would put up this matter for a decision to the head of the physics department, Professor Pegram. [2]

A telegram was sent by Weisskopf to Halban in Joliot's laboratory reminding Joliot of my letter and advising him that we were approaching the British physicists. Another telegram was sent by Weisskopf to Blackett in England suggesting that the British withhold all publications on this subject. A letter was sent by Wigner to Dirac in Cambridge, England, to the same effect. Blackett cabled to Weisskopf that the collaboration of the Royal Society could be expected, but Joliot's reply was not satisfactory. Joliot's cable pointed out that articles had appeared in the American press in February which were based on statements by [Richard B.] Roberts in Tuve's laboratory and let the cat out of the bag. To this cable of Joliot I replied that we had in the meantime secured the collaboration of Tuve's laboratory and I urged Joliot to agree to a collaboration in this matter. The answer of Joliot to my telegram was negative. *(Docs. 30–38) (Figs. 2–4)*

After Joliot's final refusal to collaborate, all my colleagues at Columbia University expressed themselves in favor of publishing our papers. I continued to take the stand that irrespective of Joliot's policy, we ought not to publish our own work. Pegram, the head of the department, was undecided. It seemed impossible to reconcile the two opposing views, and Professor I. Rabi at Columbia, who was not himself involved in this work, gave me a friendly warning that if I continued to take such an irreconcilable stand, I would probably be left without facilities for further work at Columbia. At the suggestion of Fermi, we finally agreed to leave the decision up to Pegram, the head of the department. [8] Pegram hesitated for a while to make this decision, and after a few weeks he finally came and said that he had decided that we should now publish everything.

He later told me why he decided this—and so many decisions were based on the wrong premises. Rabi was concerned about my stand because he said that everybody was opposed to withholding publication, only I wanted it in the Columbia group. This would make my position difficult, in the end impossible, and he thought that I ought to yield on this. According to what Pegram told me, Rabi had visited Urbana and found that Goldhaber in Urbana knew of our research at Columbia; from this Rabi concluded that these results were already known as far as Urbana, Illinois, and there was no point in keeping them secret. The fact was that I was in constant communication with Goldhaber; I wrote him of these results, and he was pledged to secrecy. He talked to Rabi because of course Rabi was part of the Columbia operation. So on this false premise, the decision was made that we should publish. [2] *(Docs. 41–48)*

Postal Telegraph

Received at 2708 B'WAY, N.Y.C. AC. 4-9899

N33 43 DL 4 EXTRA=PARIS FRANCE 4 VIA WB WASHINGTON DC 1235P

DR LEO SZILARD=
    DEPT OF PHYSICS COLUMBIA UNIV NYC=   APR 5 1939

BIEN RECU LETTRE SZILARD MAIS PAS CABLE ANNONCE STOP PROPOSITION DU 31 MARS TRES RAISONNABLE MAIS VIENT TROP TARD STOP AVONS APPRIS SEMAINE DERNIERE QUE SCIENCE SERVICE AVAIT INFORME PRESSE AMERICAINE 24 FEVRIER SUR TRAVAUX ROBERTS STOP LETTRE SUIT.

    JOLIO HALBAN KOWARSKY
               126P

WESTERN UNION

To: Joliot College de France Paris   Apr. 6

REPLYING YOUR CABLE WEISSKOPF CONCERNING ROBERTS PAPERS DEAL WITH DELAYED NEUTRON EMISSION WHICH IS HARMLESS ALSO MUCH WEAKER THAN HE THINKS STOP HOWEVER TUVE'S GROUP RECENTLY APPROACHED AND PROMISED COOPERATION STOP WE HAVE SO FAR DELAYED PAPERS IN VIEW OF POSSIBLE MISUSE IN EUROPE STOP KINDLY CABLE AS SOON AS POSSIBLE WHETHER YOU INCLINED SIMILARLY TO DELAY YOUR PAPERS OR WHETHER YOU THINK THAT

```
STANDARD TIME INDICATED
RECEIVED AT
2708 B'WAY, N.Y.
AC. 4-9899
TELEPHONE YOUR TELEGRAMS
TO POSTAL TELEGRAPH
```

Postal Telegraph
Mackay Radio   All America Cables
Commercial Cables   Canadian Pacific Telegraphs

THIS IS A FULL RATE TELEGRAM, CABLE-
GRAM OR RADIOGRAM UNLESS OTHERWISE
INDICATED BY SYMBOL IN THE PREAMBLE
OR IN THE ADDRESS OF THE MESSAGE.
SYMBOLS DESIGNATING SERVICE SELECTED
ARE OUTLINED IN THE COMPANY'S TARIFFS
ON HAND AT EACH OFFICE AND ON FILE WITH
REGULATORY AUTHORITIES.

Form 16

N54 14 CABLE VIA FRENCH=PARIS 1850 APR 7 1939

LC SZILARD=
    KINGSCROWN HOTEL NY=
    (SZILARD CARE KINGS CROWN HOTEL 420 W 116 ST)=

QUESTION ETUDIEE SUIS D AVIS MAINTENANT PUBLIER AMITIES=.
        JOLIOT.
            3 15P

Figures 2–4
Facsimiles of the cables exchanged between April 5–7, 1939. Figures 2 and 4 are cables from Joliot (originals in Szilard files). Figure 3 is a copy in Szilard's handwriting of the cable he sent to Joliot.

Documents from December 1938 through July 1939.

The following documents show Szilard's response to the news of the discovery of nuclear fission: a last-minute request that his patents be kept secret, inception of experiments, and immediate concern to inform physicists and other people who might help direct events.

Document 19

c/o Clarendon Laboratory
Parks Road
December 21st, 1938

Director of Navy Contracts
Admiralty
London S.W.1

Ref. No. C.P. Branch 10, Patents 8142/36

Sir:

I refer to your letter[1] of March 26th, 1936, which bears the above reference number, and beg to inform you that further experiments, carried out in collaboration with Professor M. Goldhaber, Cavendish Laboratory, Cambridge, and in cooperation with S. W. Barnes[2] of the University of Rochester, have definitely cleared up the anomalies which I have observed in 1936. This work will be published in the Physical Review.[3]

In view of this new work it does not now seem necessary to maintain the patent with which your letter deals, nor would the waiving of the secrecy of this patent serve any useful purpose. I beg therefore to suggest that the patent to withdrawn altogether.

I am, Sir,
Yours very truly,
Leo Szilard

Document 20

DIRECTOR OF NAVY CONTRACTS
ADMIRALTY
LONDON, S.W.1

JANUARY 26, 1939

REFERRING TO CP10 PATENTS 8142/36 KINDLY DISREGARD MY RECENT LETTER STOP WRITING

LEO SZILARD

---

[1]See *Scientific Papers*, pp. 733–734, for related letters.
[2]Sidney W. Barnes, physicist at Rochester.
[3]*Physical Review*, 55:47–49 (1939), in *Scientific Papers*, pp. 155–157.

Document 21

c/o Liebowitz
420 Riverside Drive
New York City
February 2nd, 1939

Director of Navy Contracts
Admirality
London S.W.1.

Ref. No. C.P. Branch 10, Patents 8142/36

Sir:

I have written to you a letter dated of December 21st, 1938, and subsequently I cancelled this letter by cabling you on January 26th as follows: "Referring to C. P. 10 patents 8142/36 kindly disregard my recent letter stop writing Leo Szilard".

I wish to explain my reason for cancelling my letter by cable, and in order to do so I have to state the following.

When I wrote you the letter of December 21st, I was satisfied that the anomalies observed in indium have been definitely cleared up, and that indium cannot be used for the process described in the patent which has been assigned by you. This is still my opinion. Since I wrote you my letter of December 21st it has however turned out that anomalies observed in another element, which appeared to be similar to the case of indium, have an entirely different explanation, and that the process underlying the anomalies of this other element might very well turn out to be similar to the process described in the patent assigned to you. Intense work is now being done on this phenomenon, and I cabled you in order to ask you not to take any action until this new work has been completed.

I should be glad if you could perhaps let me know whether my cable reached you in time to stop any action which you may have started on the basis of my preceding letter, and accordingly whether or not the status of the patent remains unchanged.[4]

Kindly note that I am at present in America and to be reached at the above address, but, of course, a letter addressed to the Clarendon Laboratory, Oxford, will also reach me with some delay.

I am, Sir,
Yours very truly,
Leo Szilard

[4]The status did remain unchanged. This later became British patent 630,726, published in *Scientific Papers*, pp. 639–651.

Document 22

Hotel King's Crown[5]
Opposite Columbia University
420 West 116th Street
New York City
January 25th, 1939

Mr. Lewis L. Strauss[6]
c/o Kuhn, Loeb & Co.
52 William Street
New York City

Dear Mr. Strauss:

I feel that I ought to let you know of a very sensational new development in nuclear physics. In a paper in the "Naturwissenschaften" Hahn reports that he finds when bombarding uranium with neutrons the uranium breaking up into two halves giving elements of about half the atomic weight of uranium. This is entirely unexpected and exciting news for the average physicist. The Department of Physics at Princeton, where I spent the last few days, was like a stirred-up ant heap.

Apart from the purely scientific interest there may be another aspect of this discovery, which so far does not seem to have caught the attention of those to whom I spoke. First of all it is obvious that the energy released in this new reaction must be very much higher than in all previously known cases. It may be 200 million [electron-]volts instead of the usual 3-10 million volts. This in itself might make it possible to produce power by means of nuclear energy, but I do not think that this possibility is very exciting, for if the energy output is only two or three times the energy input, the cost of investment would probably be too high to make the process worthwhile. Unfortunately, most of the energy is released in the form of heat and not in the form of radioactivity.

I see, however, in connection with this new discovery potential possibilities in another direction. These might lead to a large-scale production of energy and radioactive elements, unfortunately also perhaps to atomic bombs. This new discovery revives all the hopes and fears in this respect which I had in 1934 and 1935, and which I have as good as abandoned in the course of the last two years. At present I am running a high temperature and am therefore confined to my four walls, but perhaps I can tell you more about these new developments some other time. Meanwhile you may look out for a paper in "Nature" by Frisch and Meitner which will soon appear and which might give you some information about this new discovery.

With best wishes,
Yours sincerely,
Leo Szilard

---

[5]To avoid repetition, we do not print the complete address with the rest of the letters bearing it. Szilard kept it through 1941.
[6]Financier, later Atomic Energy Commissioner. See Strauss, *Men and Decisions* (Garden City, N.Y.: Doubleday, 1962), pp. 171 ff.

Document 23

CD OXFORD 153 [February 3, 1939]
SZILARD, CARE LIEBOWITZ
420 RIVERSIDE DR NY

BERYLLIUM SENT FEBRUARY 3RD REGISTERED POST[7]

LINDEMANN

Document 24

Hotel King's Crown
New York City
February 13th, 1939

Mr. Lewis L. Strauss
Brandy Rock Farm
Brandy, Va.

Dear Mr. Strauss:

I hope you and Mrs. Strauss enjoyed staying at Palm Beach and that you are now having a nice time at your farm.

After I left your train in Washington I spent a day with Dr. Teller there and another day with Dr. Wigner at Princeton and told both of them of our tentative plan to make use of the form of an "association" and let such an association take action if it seems desirable that something should be done along the lines which we discussed. Dr. Teller, who is Professor for Theoretical Chemistry at George Washington University, will be at our disposal if it becomes necessary to keep some person close to the Administration informed of the developments, and he also can get the cooperation of his colleagues in Washington if this will be required. Dr. Wigner thought that some of the experiments which we discussed could be done at Princeton. As he is an old friend of mine and has much influence in the department there is very much in favor of following his suggestion, but I feel that it will be necessary to see what the position is from the point of view of equipment, and whether some younger members of the department could cooperate without abandoning work in which they are at present engaged.

On my return to New York I went to see Fermi to tell him of all these conversations and also to discuss some of the small scale experiments which might be made in the near future.

Since my return almost every day some new information about uranium became available, and whenever I decided to do something one day it appeared foolish in the light of the new information on the next day. I found that the Radium Chemical Co. had in stock 200 mgm of radium mixed with beryllium, which is a nice constant source of fast neutrons. The rent for six months amounts to $500.00. As Fermi thought that

---

[7]Szilard received the block February 18, 1939.

he would like to use such a neutron source for his experiment I felt that I ought to get it for him. It did not seem fair to ask you to take any decisions from a distance, and so I thought it might be best that I should advance $500.00 for expenses of this type and to see later whether you could sanction the expenditures afterwards.[8] A few days later it turned out that this neutron source was too bulky to be suitable for Fermi's experiment, and Fermi said that for the present he is quite satisfied with the radon sources which he is getting anyway once a week at Columbia. In these circumstances I arranged with the Radium Chemical Co. that they will let me have one gram of radium on loan instead. This radium used in conjunction with the beryllium block sent from Oxford represents an intense source of photo-neutrons which can be used for a number of experiments. The rent is $125.00 per month, and we have to rent it for a minimum period of three months.

The outlook has changed in some important respects since I last saw you. It is now known that fast neutrons split both uranium and thorium, but slow neutrons do not split thorium, and they probably do not split the bulk of uranium either. If enough neutrons are emitted when fast neutrons split thorium or uranium it will still be necessary to see whether or not the emitted neutrons are slowed down to a velocity at which they are ineffective before they had a chance to split enough nuclei to make the maintenance of a chain reaction possible.

On the other hand, slow neutrons seem to split a uranium isotope which is present in an abundance of about 1 % in uranium.[9] If this isotope could be used for maintaining chain reactions, it would have to be separated from the bulk of uranium. This, no doubt, would be done if necessary, but it might take five or ten years before it can be done on a technical scale. Should small scale experiments show that the thorium and the bulk of uranium would not work, but the rare isotope of uranium would, we would have the task immediately to attack the question of concentrating the rare isotope of uranium.

As you see, the number of possibilities had increased since you left town. Some of the experiments which were devised, in particular the experiment which Fermi first planned, appear now to be much more difficult than before. Other experiments, such as those with photo-neutrons, are not affected, but of course they have somewhat the character of preliminary experiments.

I am enclosing a clipping which might interest you, as it shows the state of mind of some physicists on February 4th.[10] The man who inspired this article did his best to hide what he thought, but his dementi is somewhat clumsy, and he almost gives himself away in the last paragraph.

---

[8]Strauss never offered to make any such payment.
[9]i.e., uranium-235.
[10]In the *New York Times* of February 5, p. D9, Waldemar Kaempffert reported, "Hope is revived that we may yet be able to harness the energy of the atom." A "remotely possible atomic powerhouse" was depicted.

Anyway, things have calmed down to some extent in the last few days, and the newspapers at least might soon forget about uranium.

With best wishes,

Yours very sincerely,
Leo Szilard

Szilard's experiments were delayed because he did not have the connections needed to get him radium instantly:

Document 25

Radium Chemical Company, Inc.
Chicago Office
Marshall Field Annex Building
1 East 42nd Street, New York
Radium Sales Department
Radium Leasing Department
February 14, 1939

Dr. Leo Szilard
King's Crown Hotel
420 West 116th Street
New York, N.Y.

Dear Dr. Szilard:

Referring to our negotiations thus far for rental of approximately one gram of radium for physical experiments with gamma rays:

Before consummating the lease, will you be kind enough to have Dr. Wigner, or some other member of the faculty at Princeton University, state that some of the experiments will actually be conducted under the direction of the Physics Department of the University.

Our Board feels that we should have some such assurance, particularly to satisfy the insurance underwriters, of your own reliability and that the radium will have proper safeguards while out of our possession.

Yours very truly,
R. B. Kearney
Sales Manager

Meanwhile a group in Washington found neutrons produced in fission. Although these neutrons were delayed, and hence not true fission neutrons, it seemed at first that they might be able to sustain a chain reaction.

Document 26

Edward Teller
2610 Garfield Street
Washington, D.C.
[about February 17, 1939]

Dear Szilard,

Today I was out at Tuve's. Just then a letter to the Physical Review[11] was being prepared with the following contents: After bombardment of Uranium, the number of neutrons emerging with a period of 10 minutes is approximately half of that of neutrons from the fission process. This, and other experiments, show that these neutrons are not produced by X-rays.[12] These 10-minute neutrons also result from bombardment of Thorium, with the same relative yield. Thus, one may consider it certain that the neutrons are derived from a fission product.

As soon as I began taking interest in Uranium, sharp discussion started on the practical significance. Tuve, Hafstad and Roberts[13] are entirely aware of what is involved. They also know of Fermi's experiments. Of course, I didn't say anything. The above-mentioned letter cannot cause any harm. They do not plan any further experiments for lack of larger amounts of Uranium. They are particularly interested in the separation of U238 and U235 as, according to Bohr, only the latter produces fission.

I do not know their detailed plans, but I believe that urgent action is required. Very many people have discovered already what is involved. Those in Washington would like to persuade the Carnegie Institution that it should provide more money for U-research in view of the practical significance of the matter. This is perhaps not a bad idea because the Carnegie people are of good will and are cautious (perhaps a little too cautious). But right now this has no reality unless the leadership becomes more interested than it has been so far (I believe this would be desirable), due to Fermi's visit or in connection with some other occurrence.

I repeat that there is a chain-reaction mood in Washington. I only had to say "Uranium", and then could listen for two hours to their thoughts. I would be glad to hear further, and to help if possible.[14]

Teller

---

[11]Richard B. Roberts, R. C. Meyer, P. Wang, *Physical Review,* 55:510–511 (1939), submitted February 18, 1939.
[12]After fission the fragments of uranium were radioactive and emitted gamma rays; Roberts' group checked that the neutrons they observed did not result from the impact of these rays on atoms, but came directly from the fragments themselves.
[13]L. R. Hafstad and R. B. Roberts, physicists in Tuve's group at the Carnegie Institution.
[14]This letter was translated by Bela Silard for this volume from the handwritten Hungarian original and was corrected by Edward Teller.

Document 27

Hotel King's Crown
New York City
March 22nd, 1939

Dr. M. Tuve
Carnegie Institute of Terrestrial Magnetism
5241 Broad Branch Road
Washington, D.C.

Dear Tuve:

I told Fermi of the conversation which we had in Washington. I also raised the question whether he would see much advantage in your using some private funds, which I believe we could find for you if necessary, instead of the Navy funds. Since the Navy funds have already been offered to you Fermi seems to think that you might as well use those for the present. He asked me to send you a copy of his paper which you will find enclosed for your information as well as for Hafstaadt's[15] and Heydenburg's.[16]

I have telephoned to the Radium Chemical Co. and asked them if we could have another gram of radium for a minimum period of one month to be used by your group at Washington. Their reaction was so far favorable, and they cabled at my request to Brussels in order to ask for a permission.[17] I expect a telephone call from them in the next few days and will let you know as soon as I hear from them.

I have also telephoned to the Beryllium Corporation, but so far I did not succeed in getting hold of anybody to whom I could talk sense. I shall try again to-morrow. I shall keep you informed as to how things develop. At present we are trying to find out the best conditions for the "chemical" experiment.

If you want uranium oxide you can get it in ten pound packages from $2.65 to $2.70; uranium of Belgian origin from B. F. Drakenfeld, 45 Park Place, New York City, and uranium of Canadian origin from Du Pont de Nemours, The R. and H. Chemical Department, Empire State Building, New York City.

Following your suggestion that rare earth contamination might be important, we are going to test different samples for such contaminations and shall write you the results.

With kind regards to all,

Yours sincerely,
Leo Szilard

P.S. Perhaps before you definitely commit yourself to rent one gram of radium you might wish to reconsider whether you cannot do the same experiments equally well

---

[15] i.e., Hafstad.
[16] N. P. Heydenburg, physicist working with Tuve's group.
[17] The Radium Chemical Co. was a subsidiary of the Union Minière du Haut-Katanga, a Belgian firm.

with deuteron-carbon neutrons.[18] Maybe we can have another talk on this subject in a fortnight or so. By then it will be possible to see clearly what experiments are most urgently needed to complete the picture.

Document 28

>Carnegie Institution of Washington
>Department of Terrestrial Magnetism
>5241 Branch Road, N.W.
>Washington, D.C.
>Monday 3/27/39

Dear Szilard

Thanks for your two letters, + MS of Fermi's expts.

We are in the midst of an expt to demonstrate simultaneous neutrons > 500 kv similar to your photo-n[eutron] expt, but using C[arbon]-neutrons. This type of expt gives very little of quantitative value, but after 2 or 3 more days I will stop it + run up to NY to talk to you + Fermi.

Do not cable for Be block as yet.[19] Thanks.

President Bush (CIW)[20] insisted that we stay clear of the Navy; we can have what funds we need, after we decide what really is worth doing. At present I think the Fermi expt with large (200 lbs) amounts of UO_2 is the only good expt. We could use our water tank 8 ft diam by 8 ft deep, and with ample neutrons 100 to 1000 curies (new machine) make a real measure of area under the "tail."[21] Before making any moves or purchases we will confer with you + Fermi.

Regards—Tuve

[Transcribed from handwritten letter with marginal note]

Please advise Fermi.
Joliot's expt 6×10^{-25} hardly agrees with his interpretation of > 1 n[eutron] per fission?[22]

The following group of letters illustrates the efforts Szilard and others made to encourage self-censorship of fission publication. Szilard's chief concern from the start was Joliot, whom he suspected to be doing fission experiments.

---

[18] Produced by bombarding carbon with deuterons from an accelerator.
[19] Szilard had offered the use of a beryllium block so Tuve's group could have a radium-beryllium neutron source.
[20] Vannevar Bush, then president of the Carnegie Institution of Washington.
[21] i.e., the "tail" or outer end of a curve of neutron density in a tank filled with uranium oxide and water. Measuring this would indicate how close the system was to a chain reaction.
[22] Refers to H.v.Halban, Jr., F. Joliot, L. Kowarski, *Nature, 143*: 470–471 (1939). 6×10^{-25} was suggested as the cross-section (probability) for producing a neutron in fission, but the French noted that this was a lower limit "certainly inferior to the actual value."

Document 29

c/o Liebowitz
420 Riverside Drive
New York City
February 2nd, 1939

Professor F. Joliot
Laboratoire de Chemie Nucléaire[23]
Collège de France
Paris

Dear Professor Joliot:

The only reason for my writing to you this letter to-day is the remote possibility that I shall have to send you a cable in some weeks, and if that happens this letter will help you to understand what the cable is about. This letter is therefore merely a precaution, and we hope an unnecessary precaution.

When Hahn's paper reached this country about a fortnight ago, a few of us got at once interested in the question whether neutrons are liberated in the disintegration of uranium. Obviously, if more than one neutron were liberated, a sort of chain reaction would be possible. In certain circumstances this might then lead to the construction of bombs which would be extremely dangerous in general and particularly in the hands of certain governments.

It is of course not possible to prevent physicists from discussing these things among themselves, and, as a matter of fact, the subject is fairly widely discussed here. However, so far, every individual exercised sufficient discretion to prevent a leakage of these ideas into the newspapers.

In the last few days there was some discussion here among physicists whether or not we should take action to prevent anything along this line from being published in scientific periodicals in this country, and also ask colleagues in England and France to consider taking similar action. No definite conclusions have so far been reached in these discussions, but if and when definite steps are being taken I shall send you a cable to tell you what is being done.

We all hope that there will be no, or at least not sufficient, neutron emission and therefore nothing to worry about. Still, in order to be on the safe side, efforts are made to clear up this point as quickly as possible. Experiments at Columbia University are in charge of Fermi and will perhaps be the first to give reliable results.

Perhaps you have also thought of the same things and have contemplated or started such experiments. Maybe you are able to get definite results at an earlier date, which, of course, would be very valuable help towards ending the present disquieting uncertainty. Whatever information on the subject you might care to transmit by letter or cable at some later date will, I am sure, be greatly appreciated. Also, should you

---
[23] i.e., Prof. F. Joliot, Lab. de Chimie Nucléaire.

come to the conclusion that publication of certain matters should be prevented, your opinion will certainly be given very serious consideration in this country.[24]

Yours sincerely,
Leo Szilard

Document 30

LEWIS L. STRAUSS [February 19, 1939]

PRIVATE COMMUNICATION JUST RECEIVED[25] SHOWS THAT NEUTRONS ARE EMITTED IF URANIUM IS SPLIT BY VERY FAST NEUTRONS STOP EASY DETECTION IS DUE TO TIME LAG OF 10 SECONDS IN EMISSION STOP DOES NOT MEAN THAT OUR PROBLEM IS SOLVED IN POSITIVE SENSE STOP NUMBER AND VELOCITY OF EMITTED NEUTRONS NOT KNOWN STOP YET RESULT HAS BEARING ON OUR PROBLEM STOP IF WE WISH TO PREVENT PUBLICATION WOULD HAVE TO ACT FAST THROUGH ADMINISTRATION AND EVEN SO SUCCESS DOUBTFUL STOP DR. WIGNER IN TOWN TUESDAY AND DR. TELLER WEDNESDAY

SZILARD

The following telegrams and letters were provoked by the French publication revealing that some neutrons are produced in fission—a key fact:

Document 31

BLACKETT [Weisskopf, March 31, 1939]
PHYSICS DEPARTMENT
VICTORIA UNIVERSITY
MANCHESTER

PHYSICISTS HERE HAVE SENT PAPERS TO PHYSICAL REVIEW ON SUBJECT RELATED TO HALBAN JOLIOT LETTER TO NATURE STOP AUTHORS AGREED TO DELAY PUBLICATION IN VIEW OF REMOTE BUT NOT NEGLIGIBLE CHANCE OF GRAVE MISUSE IN EUROPE STOP IT IS SUGGESTED THAT PAPERS BE SENT TO PERIODICALS AS USUAL BUT PRINTING BE DELAYED UNTIL IT IS CERTAIN THAT NO HARMFUL CONSEQUENCES TO BE FEARED STOP RESULTS WOULD BE COMMUNICATED IN MANUSCRIPTS TO COOPERATING LABORATORIES IN AMERICA ENGLAND FRANCE AND DENMARK STOP IS IT POSSIBLE FOR YOU TO OBTAIN COOPERATION OF NATURE AND PROCEEDINGS?[26] WIGNER WRITING DIRAC STOP WEISSKOPF[27] FINE HALL PRINCETON NJ

---

[24]Szilard crossed out a concluding remark: "In the hope that there will not be sufficient neutrons emitted by uranium, I am. . . ."
[25]See letter from Teller, *Doc. 26.*
[26]i.e., the periodicals *Nature* and the *Proceedings of the Royal Society of London.*
[27]Victor Weisskopf, then physicist at the University of Rochester, who at this time was visiting Princeton; his telegrams were written in close consultation with Szilard.

Document 32

[Weisskopf, March 31, 1939]

HANS VON HALBAN[28]
11 RUE GUYNEMER
SCEAUX SEINE

KINDLY INFORM JOLIOT THAT PAPERS RELATING TO SUBJECT OF YOUR JOINT NOTE TO NATURE HAVE BEEN SENT BY VARIOUS PHYSICISTS TO PHYSICAL REVIEW BEFORE PUBLICATION OF YOUR NOTE STOP AUTHORS AGREED HOWEVER TO DELAY PUBLICATION FOR REASONS INDICATED IN SZILARDS LETTER TO JOLIOT FEBRUARY SECOND AND THESE PAPERS ARE STILL HELD UP STOP NEWS FROM JOLIOT WHETHER HE IS WILLING SIMILARLY TO DELAY PUBLICATION OF RESULTS UNTIL FURTHER NOTICE WOULD BE WELCOME STOP IT IS SUGGESTED THAT PAPERS BE SENT TO PERIODICALS AS USUAL BUT PRINTING BE DELAYED UNTIL IT IS CERTAIN THAT NO HARMFUL CONSEQUENCES TO BE FEARED STOP RESULTS WOULD BE COMMUNICATED IN MANUSCRIPTS TO COOPERATING LABORATORIES IN AMERICA ENGLAND FRANCE AND DENMARK STOP COMMUNICATING BLACKETT AND DIRAC IN ATTEMPT TO GET COOPERATION OF NATURE AND PROCEEDINGS ROYAL SOCIETY STOP PLEASE CABLE WEISSKOPF FINE HALL PRINCETON NJ

Document 33

[Wigner]
Palmer Physical Laboratory
Princeton University
Princeton New Jersey
March 30, 1939

Dear Paul [Dirac]:[29]

I am writing to you in a rather serious matter this time. The enclosed letter, sent by Szilard to Joliot on February 2nd is self explanatory. Experiments undertaken since that time by Fermi and by Szilard did not help to dispel the fear which prompted Szilard's letter. In realisation of the danger mentioned in this letter, all efforts are made here to delay publications relating to this subject as these could possibly enhance the danger of a grave misuse by certain powers. The papers of Szilard and of Fermi, although received by the Physical Review some time ago, are withheld from publication and it is intended that they be printed only in the form of reprints to be distributed among the most interested laboratories in England, the U.S., France and Denmark. Similar arrangements are intended for all papers on this subject by other workers in the United States.

Halban-Joliot-Kowarski's letter to Nature prompted the physicists who loyally cooperated here to inquire today by cable concerning Joliot's attitude in this matter.

---

[28]Hans von Halban, Jr., coworker of Joliot and old friend of Weisskopf.
[29]P. A. M. Dirac, physicist at Cambridge.

Bohr undertakes to communicate with Copenhagen and a cable is sent simultaneously to Blackett. The proposition made in these communications is to use for the publication of all papers, relating to this subject, the method foreseen for this purpose for workers in the U.S. and described above.

What we would like to ask you at this time is to get in touch with Blackett and to actively support him in his endeavours if you find our position to be the reasonable one.

It is my impression that there is some urgency in the matter. Although there exists apparently a great willingness for cooperation here, it is realised that the interests of the scientific workers in the U.S. may be prejudiced to some extent if America abeyed alone by the proposed procedure.

Hoping to hear from you soon and with best regards to all,

Sincerely,
Jenö [Wigner]

Document 34

Victor Weisskopf
University of Rochester
[about April 1, 1939]

Dear Blackett,

I hope you were not too much upset about my telegram but I believe that you realize the great danger which would arise, if one really could construct a bomb with uranium. The probability that this is possible might be small, but the product of the probability with the graveness of the consequences is high.

In enclose here first a letter which Szilard has written to Joliot Febr. 2. Joliot has not answered this letter and we do not know Joliot's attitude to the whole situation after his recent publication. I have sent to Halban a similar telegram as to you urging him to cooperate.[30]

Further I enclose a note which Szilard has sent to Physical Review but the publication of which is being delayed. There are other papers from Columbia sent in and kept back, which could be sent to you if the cooperation begins to work. I am also enclosing a letter from Szilard to myself which gives you further details about his experiments.

I would like to tell you how far the cooperation here for delaying "dangerous" manuscripts has developed so far. We know that the group around Tuve is now willing to cooperate. Lawrence is coming here on April 3rd. and we shall discuss the matter

---

[30]See *Doc. 29.*

with him then. Tate[31] (editor of Phys. Rev.) is being approached and it is suggested that authors who may send in manuscripts concerning "dangerous" neutron emissions be advised to communicate with us. We shall send you a cable when a definite procedure has been decided upon in connection with Phys. Rev.

Much love to your family,
Very truly yours,
[Weisskopf]

The French rejected Szilard's offer while the British accepted it:

Document 35

DR LEO SZILARD [April 5, 1939]
DEPT OF PHYSICS COLUMBIA UNIV NYC

SZILARD LETTER RECEIVED BUT NOT PROMISED CABLE STOP PROPOSITION OF MARCH 31 VERY REASONABLE BUT COMES TOO LATE STOP LEARNED LAST WEEK THAT SCIENCE SERVICE HAD INFORMED AMERICAN PRESS FEBRUARY 24 ABOUT ROBERTS' WORK[32] STOP LETTER FOLLOWS[33]

JOLIOT HALBAN KOWARSKI[34]

Document 36

JOLIOT [April 6, 1939]
COLLEGE DE FRANCE PARIS

REPLYING YOUR CABLE WEISSKOPF STOP ROBERTS PAPERS CONCERNING DELAYED NEUTRON EMISSION WHICH IS MUCH WEAKER THAN HE THINKS AND HARMLESS STOP HOWEVER TUVES GROUP WAS RECENTLY APPROACHED AND PROMISED COOPERATION STOP WE HAVE SO FAR DELAYED PAPERS IN VIEW OF POSSIBLE MISUSE IN EUROPE STOP KINDLY CABLE AS SOON AS POSSIBLE WHETHER INCLINED SIMILARLY TO DELAY YOUR PAPERS OR WHETHER YOU THINK THAT WE SHOULD NOW PUBLISH EVERYTHING[35] STOP

KINGS CROWN HOTEL SZILARD

---

[31]John T. Tate, University of Minnesota physicist, editor of *Physical Review*.
[32]Science Service releases of March 24 and April 16, 1939, erroneously reported Roberts had found neutron production which could make chain reactions feasible.
[33]Our translation. The French original is as follows:
BIEN RECU LETTRE SZILARD MAIS PAS CABLE ANNONCE STOP PROPOSITION DU 31 MARS TRES RAISONNABLE MAIS VIENT TROP TRAD STOP AVONS APPRIS SEMAINE DERNIERE QUE SCIENCE SERVICE AVAIT INFORME PRESSE AMERICAINE 24 FEVRIER SUR TRAVAUX ROBERTS STOP LETTRE SUIT
JOLIOT HALBAN KOWARSKY
[34]Lew Kowarski, coworker of Halban and Joliot
[35]In his first draft, not sent, Szilard added: "Regardless of possible consequences."

Document 37

King's Crown Hotel
New York City
Apr. 7—39.

Dear Professor Joliot,

Enclosed you will find a manuscript for your personal information, which has been sent to Phys. Rev. on March 16th.—Its publication is being delayed for reasons of which you know. A definite policy in this respect was formulated on March 20th and I was on the point of cabling you on that day when I was told about your note to Nature.—I have cabled to you yesterday in reply to your cable to Weisskopf (who returned to Rochester N.Y.) and when we shall have your cable reply the matter of withholding publications may come up again for discussion.

With best wishes,
Yours sincerely,
Leo Szilard

Document 38

[April 6, 1939]

QUESTION STUDIED MY OPINION IS TO PUBLISH NOW REGARDS JOLIOT[36]

Document 39

WEISSKOPF [April 8, 1939]
FINE HALL PRINCETON (NJ) USA

YOUR SUGGESTION PASSED TO NATURE AND ROYAL WHO WILL SURELY COOPERATE STOP AWAITING LETTER WITH DETAILS

BLACKETT

Document 40

Hotel King's Crown
New York City
April 11th, 1939

Mr. Lewis L. Strauss
25 East 76th Street
New York City

Dear Mr. Strauss:

These are just a few lines to keep you informed of how things have developed since I last saw you. Fermi and I were sorry that we could not see you in Virginia when we were in Washington.

---

[36]Our translation. The French original reads:
QUESTION ETUDIEE SUIS D'AVIS MAINTENANT PUBLIER AMITIES JOLIOT.

The following is for your private information only. Cooperation was established in Washington with the Carnegie Institute for Terrestrial Magnetism,[37] and also contacts were made through the official channels via the Physics Department of Columbia University with the Navy. These contacts are perhaps too loose, but for the present this is of no importance.

Since my return from Washington I cut down all extra-laboratory activities and tried to get more information about the number of neutrons emitted, which is the most important point at present. Though this number seems to be above one, I am still not certain about it. Fermi bases his plans on the belief that the number is larger than one.

Accordingly we are preparing an experiment on a semi-large scale, using 500 pounds of uranium oxide. I am glad to say that we could borrow this amount, otherwise I might have approached you for financial assistance.

So far publication of the papers, which were sent to the Physical Review on March 16th, is being delayed at our request, and efforts are made to get similar action in England and France. In the mean time a paper by Joliot appeared in Nature, which relates to our subject, but so far it did not attract much attention. Now we are trying to get Joliot to co-operate, but I do not know whether we will succeed.

Some time ago Loomis asked Fermi out to Tuxedo Park, and I understand from Richards that Loomis talked to you over the telephone after Fermi's visit. Since then Loomis once inquired over the telephone, asking Fermi about the present state of the experiments.

I hope to see you some time when you are not very busy, and in any case I shall let you know of the further developments.

With best wishes,

Yours sincerely,
Leo Szilard

Document 41

[Goldhaber]
University of Illinois
Department of Physics
Urbana
April 12, 1939

Dear Szilard,

Thank you for the paper. It reached me just before I was leaving for a trip to Madison, Wisconsin, together with Rabi.

I suppose you would have detected delayed neutrons with the He[lium] chamber if these neutrons had an energy of $\sim .5$ Mev as Roberts et al. estimate. This makes either Roberts' estimate of the cross section for production of the 'delayed neutron' emitter much too large or your multiplication value too small.[38] Roberts' estimate seems on

---

[37] i.e., Tuve's group at the Department of Terrestrial Magnetism of the Carnegie Institution.
[38] Szilard's value was about right.

the face of it unlikely high.—Have you any idea how fast your multiplication neutrons are?

Of course, it is a good idea to keep results secret until it is settled whether the U-bomb is possible, in reasonable dimensions, or not. Let us hope not. But if yes, it is important to be a step ahead of the dictator countries, though I fear, it means only a step ahead. Have you thought of definite steps of buying off all U (&Th??) supplies?[39] A map of their distribution on the earth looks a little discouraging, though it is only a qualitative one, which I looked up.—I fear you will not be too successful with the attempt to keep results secret. The first indications that people in Berlin are doing similar work appear in Naturwiss. (arrived to-day) in an article of Droste, who with Reddeman is looking for faster neutrons, when D + D neutrons are used on U.[40] He mentions this at the end of a letter. As soon as the first papers from Germany appear, I am sure, many of those who have agreed to secrecy will see no further good in it, and though they are right as far as the ultimate results are concerned, the immediate effect of secrecy is very important. Have you any results on Th?

Talking of 'secrets': Please keep this a secret for the time being. I have become engaged to Trude Scharff, and working on the hypothesis that there will be no war within the next few weeks (a very weak hypothesis, I fear) I have sent her an invitation (from Loomis) to come here on a visitor's visa. When she is here we can get married and then she can immigrate non-quota via Canada. This seems the only workable plan in a hurry.

I have written to Fowler.[41]

Kind regards,
Yours, M. Goldhaber

When Pegram decided that the Columbia scientists' fission papers should be published, Szilard had to inform his correspondents of his failure and the reasons for it:

Document 42

PROFESSOR BLACKETT [April 14, 1939]
VICTORIA UNIVERSITY
MANCHESTER

ACTION PROPOSED WEISSKOPFS CABLE DROPPED HERE YESTERDAY AND PAPERS WILL APPEAR PHYSICAL REVIEW STOP LETTER FOLLOWS SZILARD

---

[39]I.e., Uranium and Thorium.
[40]G. von Droste and H. Reddeman, *Naturwissenschaften 27*: 371-372 (1939), similar to the work of Szilard and Zinn, using neutrons produced by bombardment with deuterons.
[41]Probably William Fowler, nuclear physicist at the California Institute of Technology.

Document 43

Hotel King's Crown
New York City
April 14th, 1939

Dear Blackett:

Referring to your cable which I received from Weisskopf I have to inform you of the following: A cable correspondence with Joliot showed that Joliot is not inclined to delay his future papers and apparently is of the opinion that it is too late for the proposed action which might have been a reasonable solution if it had been applied at an earlier date. Influenced by Joliot's stand and also by papers which have been printed in the C. R.,[42] it has been decided here to publish the papers which have been sent to Physical Review some time ago and which were so far held up. A cable will be sent to you today or to-morrow to inform you that no action along the lines suggested by Weisskopf will at present be pursued in this country. I personally regret this decision for reasons which I have to explain to you some other time, this being a hurried note only. It is conceivable that there will be a change of attitude, and if that happens undoubtedly you will be informed as soon as possible.

With best wishes,
Yours sincerely,
Leo Szilard

Document 44

King's Crown Hotel
New York City
April 14th, 1939

Dear Mr. Strauss:

After an exchange of cables with Joliot in Paris and Blackett in England the Physics Department at Columbia University decided to publish our papers which were sent to Physical Review some time ago. This decision which runs contrary to my personal wishes was largely based on Joliot's unwillingness to delay his papers in connection with his view that the situation has already got out of hand.

I am enclosing a manuscript of the paper which will appear in the next issue of Physics Review.[43]

With best wishes,

Yours sincerely,
Leo Szilard

[42]*Comptes Rendus de l'Académie des Sciences*, Paris.
[43]*Scientific Papers*, p. 158 (paper submitted March 16 and published April 15, 1939).

Document 45

Lewis L. Strauss
52 William Street
New York
April Seventeenth, 1939

Dear Dr. Szilard,

It was good of you to send me the very interesting manuscript of your paper which I have read with the greatest interest. I share your regret that it was not possible to defer publication of these developments at this more or less critical juncture.

Hoping to see you within the very near future, I am, as ever,

Faithfully yours,
Lewis Strauss

Dr. Leo Szilard
King's Crown Hotel
420 West 116th Street
New York, N.Y.

Document 46

JOLIOT [about April 14, 1939]
COLLEGE DE FRANCE
PARIS

HAVING RECEIVED YOUR CABLE IT HAS BEEN DECIDED TO PUBLISH PAPERS PREVIOUSLY SENT TO PHYSICAL REVIEW GREETINGS

SZILARD[44]

Document 47

[Joliot]
Collège de France
Laboratoire de Chimie Nucléaire
Place Marcelin-Berthelot
Paris (V e)
Tel.: Odéon 81-60
Paris, April 19, 1939

Monsieur L. Szilard
Kings Crown Hotel
420 West 116th Street
New York

Dear Szilard,

I received your letter of April 7 and your interesting note on the liberation of neutrons. We have continued research on this question and you will find herewith the manu-

---

[44]It is not certain whether this telegram was sent, since no copy exists among the Joliot papers, Radium Institute, Paris.

script text of a letter we sent to Nature. Unfortunately it is too late for us to add your communication as reference; however, we shall certainly do this in a general article to be published in the near future.

I was very troubled about the question of postponing publication on this subject, for I am certainly among the first to understand your reasons. But you can understand that we, as well as others you may have warned, are not the only ones to be concerned with the question and we soon saw articles in scientific publications and the newspapers, in France and abroad, in which the energy consequences of the phenomenon were clearly explained. These are the only reasons for the wording of my last cable. I certainly agree with the principle of an entente, but if it is to be effective it must be extended to *all* the laboratories which could concern themselves with this question.

I would appreciate your informing your American colleagues of these considerations.[45]

Sincere greetings,
F. Joliot

[French original of the preceding letter]

Mon cher Szilard,

J'ai bien reçu votre lettre du 7 avril et votre intéressante note sur la libération des neutrons. Nous avons continué les recherches sur cette question et vous trouverez ci-joint le texte manuscrit d'une note que nous avons envoyé à Nature. Il est malheureusement trop tard pour que nous puissions ajouter en référence votre communication, cependant nous ne manquerons pas de le faire dans un article général qui sera publié prochainement.

J'étais très embarrassé en ce qui concerne l'ajournement des publications sur ce sujet, étant certainement l'un des premiers à comprendre vos raisons. Cependant vous pouvez comprendre que nous ne sommes pas, ainsi que ceux que vous avez pu prévenir, les seuls à nous occuper de cette question, et rapidement nous avons pu lire dans des publications scientifiques et dans la presse d'information, en France et à l'étranger, des articles où étaient clairement expliquées les conséquences énergétiques du phénomène en question. Ce sont les seules raisons qui ont motivé les termes de mon dernier câble. Je suis certainement d'accord avec le principe d'une entente, mais pour qu'elle soit efficace il faut qu'elle soit étendue à *tous* les laboratoires susceptibles de s'occuper de la question.

Je vous serais reconnaissant de bien vouloir faire part de ces considérations aux collègues américains que vous avez pu toucher.

Avec mes sincères salutations,
F. Joliot

[45]Our translation.

Document 48

Professor F. Joliot
Collège de France
Paris

July 5th, 1939

Dear Professor Joliot:

I thank you for your letter of April 19th and the enclosed manuscript. Please excuse my delay in acknowledging its receipt. I personally regret very much that it was not possible to arrange for a universally accepted policy with regard to publications. As far as I can see, the really dangerous questions have so far not been raised, and maybe it will be possible later, if the necessity arises, to halt certain types of publications and so to establish a sort of second line of defense. I may write to you again about this at some later date.

    Meanwhile I am, with kind regards to all,

Yours very sincerely,
Leo Szilard

Chapter III
*Sir: Some recent work by E. Fermi and
L. Szilard... leads me to expect....
Yours very truly, A. Einstein.*

In the following months Fermi and I teamed up in order to explore whether a uranium-water system would be capable of sustaining a chain reaction.[1] The experiment was actually done jointly by Herbert Anderson, Fermi, and I. We worked very hard at this experiment and we saw that under the conditions of this experiment more neutrons are emitted by uranium than absorbed by uranium. We were inclined to conclude that this meant that the water-uranium system would sustain a chain reaction. Whether finally we should have said that in print I do not know. However, the fact is that we believed it until George Placzek dropped in for a visit.[2] Placzek said that our conclusion was wrong because in order to make a chain reaction go, we would have to eliminate the absorption of [neutrons by the] water; that is, we would have to reduce the amount of water in the system, and if we reduced the water in the system, we would increase the parasitic absorption of [neutrons by] uranium. He recommended that we abandon the water-uranium system and use helium for slowing down the neutrons. To Fermi this sounded funny, and Fermi referred to helium thereafter invariably as Placzek's helium.

I took Placzek a little more seriously, and while I had no enthusiasm for helium, for purely practical reasons, I dropped then and there my pursuit of the water-uranium system. While Fermi went on examining this system in detail and trying to see whether by changing the arrangements he could not improve it to the point where it would sustain a chain reaction, I started to think about the possibility of using perhaps graphite instead of water. This brought us to the end of June. We wrote up a paper,[3] Fermi left for the summer to go to Ann Arbor, and I was left alone in New York.[4] I still had no position at Columbia; my three months were up,[5] but there were no experiments going on anyway and all I had to do was to think.

In the summer of 1939, in July, I made the first beginnings in computing the uranium-graphite system. As soon as I saw that the uranium-graphite system might work, I wrote a number of letters to Fermi telling him that I felt that this was a matter of some urgency; we should not waste our time by making detailed physical measurements of the individual constants involved, but rather try to get a sufficient amount of

---

[1]The water—or some other "moderator"—is necessary to slow down the fast neutrons produced in fission, since only slow neutrons will readily provoke further fissions.
[2]George Placzek, an emigré physicist at Cornell University.
[3]H. L. Anderson, E. Fermi, and L. Szilard, *Physical Review, 56*:284–286 (1939), in *Scientific Papers*, pp. 160–162.
[4]Fermi dropped fission work and attended the University of Michigan's summer school for theoretical physics.
[5]As a guest, March 1–May 31, 1939.

graphite and uranium to approach the critical mass and build up a chain-reacting system. Fermi's response to this crash program was very cool.[6] He said that he had thought of the possibility of using carbon instead of water, that he had computed how a homogeneous mixture of carbon and uranium would behave, and that he found that the absorption of carbon would have to be indeed exceedingly low in order to make such a system chain-reacting. I knew very well that Fermi must be aware of the fact that a homogeneous mixture of uranium and carbon is not as good as a heterogeneous uranium-carbon system; he computed the homogeneous mixture only because it was the easiest to compute.[7] This showed me that Fermi did not take this matter really seriously. This was one of the factors which induced me to approach the government quite independently of whatever Fermi or Columbia University might feel.

In July, when I reported to Pegram my optimistic views about graphite and told him why I thought the matter was urgent, he took the position that even though the matter appeared to be rather urgent, this being summer and Fermi being away there was really nothing that usefully could be done until the fall—September or perhaps October. This was the second factor which induced me to disregard everything else and go to the government directly.[8] (*Docs. 49–51*)

In July, after I took a rather optimistic view of the possibility of setting up a chain reaction in graphite and uranium, I approached Ross Gunn and told him that the situation did not look too bad; that the situation, as a matter of fact, looked so good that we ought to experiment at a faster rate than we had done before; that we had no money for this purpose; and I wondered whether the Navy could make any funds available. Afterwards I had a letter in reply, in which Ross Gunn explained that there was almost no way in which the Navy could support this type of research, but that if we got any results which might be of interest to the Navy, they would appreciate it if we would keep them informed. (*Doc. 52*) This was the second approach to the government.[9]

When Wigner came to New York I showed him [the calculations of a graphite-uranium system] I had done. He was impressed and he was concerned. At this point both Wigner and I began to worry about what would happen if the Germans got hold of large quantities of the uranium which the Belgians were mining in the Congo. So we began to think, through what channels could we approach the Belgian government and warn them against selling any uranium to Germany?

At this point it occurred to me that Einstein knew quite well the Queen of the Belgians, and so I suggested [to Wigner] that we visit Einstein, tell him about the situation, and ask him whether he might not write to the Queen. We knew that Einstein

---

[6] Szilard-Fermi correspondence, July 3–11, 1939, in *Scientific Papers*, pp. 193–198.
[7] In a "heterogeneous" mixture lumps of uranium would be distributed through the carbon in a lattice. Szilard was starting to try to calculate the possibilities of such systems.
[8] On the original tape this paragraph and the preceding paragraph come later, out of chronological order. See n. 5 to chapter IV.
[9] In the original tapes this paragraph came later, out of chronological order. See n. 12 in this chapter. If the Navy was disinterested, this was not from lack of effort on Gunn's part.

was somewhere on Long Island but we didn't know precisely where, so I phoned his Princeton office and I was told he was staying at Dr. Moore's cabin at Peconic, Long Island. Wigner had a car and we drove out there and tried to find Dr. Moore's cabin. We drove around for a half an hour, asking everybody we met—no one knew Dr. Moore's cabin. We were on the point of giving up and going back to New York when I saw a boy aged maybe seven or eight standing on the curb. I leaned out of the window and I said, "Say, do you by any chance know where Professor Einstein lives?" The boy knew that and he offered to take us there, though he had never heard of Dr. Moore's cabin,

We discussed the situation with Einstein. This was the first Einstein had heard about the possibility of a chain reaction. He was very quick to see the implications and perfectly willing to do anything that needed to be done. [2] He was willing to assume responsibility for sounding the alarm even though it was quite possible that the alarm might prove to be a false alarm. The one thing that most scientists are really afraid of is to make a fool of themselves. Einstein was free from such a fear and this above all is what made his position unique on this occation. [9]

He was reluctant to write to the Queen of the Belgians, but he thought he would write to one of the members of the Belgian cabinet whom he knew. He was about to do that when Wigner said that we should not approach a foreign government without giving the State Department an opportunity to object. So Wigner proposed that we write the letter and send a copy to the State Department with a covering letter. Einstein should say in that covering letter that if we didn't hear within two weeks from the State Department then we would send off the letter to Belgium. This is where we left off when we returned to New York and Wigner left for California.[10]

---

[10]The Szilard files contain the first draft of a letter to the Ambassador of Belgium, dictated by Einstein in German and taken down in longhand by Wigner. This draft, which Szilard subsequently translated into English, stated that Einstein wanted to draw the Ambassador's attention to discoveries which might affect the welfare of his nation and others. It appeared not only possible but highly likely that it would be possible to make a powerful source of energy from the element uranium, whose chief source was the Belgian Congo. This was indicated by recent scientific publications and by unpublished work done at Columbia University. There was a possibility that explosive bombs of unimaginable power could be constructed of uranium. If this became known, presumably certain powers would attempt to secure large stocks of the element or its ore. It seemed necessary, the draft concluded, to take precautions to keep stocks out of the hands of potential enemies. Germany, which had offered uranium for sale after taking over a mine in Czechoslovakia, was no longer allowing exports of the material. (Szilard made several changes from the German Einstein-Wigner draft, writing it for Einstein's signature alone rather than the group's and toning down references to the likelihood of a war use for uranium.)

At the same time Szilard drafted an accompanying letter in English for Einstein's signature to the Secretary of State in Washington, D.C. This said that Einstein was enclosing a draft of a letter which he had thought of sending to the Belgian Ambassador, believing that it was in the interests of this country as well as in the interest of Belgium to draw the attention of the Belgian government to a potential danger which had arisen in connection with the new development in physics. He was informing the Secretary of State in order to ask whether his department would care to receive information on this subject and to approach the Belgian government, or whether Einstein should inform the Belgian government directly through its ambassador.

Szilard submitted these drafts to Professor Einstein but, as the following shows, their tactics were changed and these letters were never sent.

This story shows that we were all green. We did not know our way around in America, we did not know how to do business, and we certainly did not know how to deal with the government. However, I had an uneasy feeling about this approach and I felt I would need some help, I would need to talk to somebody who knew a little bit better how things were done. I knew Dr. Gustav Stolper. He used to live in Berlin, he was the publisher of a leading German economic journal, and a member of the German parliament; now he was living as a refugee in New York. I went to him and talked the situation over with him. He said that he thought that Dr. Alexander Sachs, who was economic advisor to the Lehman Corporation and who had previously worked for the New Deal, might be able to give us advice on how to approach the American government—whether we should approach the State Department or some other agency of the government. He telephoned Dr. Sachs and I went to see him and I told him my story. Sachs said that if Einstein were to write a letter he would personally deliver it to President Roosevelt, and that he thought that it was no use going to any of the agencies, the War Department, or any of the government departments, because they would not know how to handle this. This should go to the White House. This sounded like good advice, and I intended to follow it. (*Doc. 53*)

In the meantime Teller had arrived in New York and I asked Teller to drive me out to Long Island (he had a car). Teller and I went to see Einstein and discussed with Einstein this new approach, the idea of writing to the President. Einstein was perfectly willing to do this. We discussed a little bit what should be in this letter. I said I would draft it and send Einstein perhaps one or two drafts for him to choose.[11] [2]

I prepared a long draft and a short draft. We did not know just how many words one could put in a letter which a President is supposed to read. How many pages does the fission of uranium rate? So I sent Einstein a short version and a longer version; Einstein thought the longer one was better, and this is the version which he signed. The letter was dated August 2, 1939. I handed it to Dr. Sachs for delivery to the White House.[12] (*Docs. 54-58*)

August passed and nothing happened. September passed and nothing happened. Finally I got together with Teller and Wigner and we decided we'd give Sachs two more weeks, and if nothing happened we would use some other channel to the White House. (*Docs. 59-63*) However, suddenly Sachs began to bestir himself, and we received a phone call from him in October saying that he had seen the President and transmitted Einstein's letter to him, and that the President had appointed a committee under the chairmanship of Lyman J. Briggs, director of the National Bureau of Standards. (*Doc. 64*) Other members of the committee were Colonel Adamson of the Army and Hoover from the Navy.[13] The committee was to meet on October 21st, and

---

[11]On this visit Einstein dictated another draft in German which Szilard took down in longhand, this time for the President of the United States. See headnote to *Doc. 54*.
[12]In the original tapes there follow two paragraphs that have been put elsewhere to preserve chronological order; see n. 9 in this chapter and n.10 in chapter II, Recollections.
[13]Col. Keith R. Adamson, Army Ordnance Department, and Commander Gilbert C. Hoover, Navy Bureau of Ordnance.

Briggs wanted to know who else he should include. I told Sachs that apart from Wigner and me, I thought that Edward Teller ought to be invited because he lived in Washington and he could act as a liaison between us and the committee. This was done. In addition, Briggs invited Dr. Tuve. Dr. Tuve had to go to New York and so he suggested that Dr. Roberts sit in for him.

It was our general intention not to ask the government for money, but to ask only for the blessing of the government, so that then with that blessing we would go to foundations, raise the funds, and get some coordinated effort going.[14] However, these meetings never go the way you have planned them.

After I presented the case and Wigner spoke, Teller spoke; and Teller spoke in two capacities. In his own name he strongly supported what I had said and what Wigner had said. Then he said that, having spoken for himself, he would now speak for Dr. Tuve. Dr. Tuve could not attend the meeting, but he had visited New York and he had a discussion with Fermi. It was Dr. Tuve's opinion that at this time it would not be advisable—"No," said Teller, "that's not what he said"—it would not be possible to spend more money on this research than $15,000.

We had no intention to ask for any money from the government at this point, but since the issue of money was injected, the representative of the Army asked, "How much money do you need?" And I said that all we need money for at this time is to buy some graphite, and the amount of graphite we would have to buy would cost about $2,000. Maybe a few experiments which would follow would raise the sum to $6,000—it was this order of magnitude.

At this point the representative of the Army started a rather longish tirade. He told us that it was naïve to believe that we could make a significant contribution to defense by creating a new explosive. He said that if a new weapon is created, it usually takes two wars before one can know whether the weapon is any good or not. Then he explained rather laboriously that it is in the end not weapons which win the wars, but the morale of the troops. He went on in this vein for a long time until suddenly Wigner, the most polite of us, interrupted him. He said in his high-pitched voice that it was very interesting for him to hear this. He always thought that weapons were very important and that this is what costs money, and this is why the Army needs such a large appropriation. But he was very interested to hear that he was wrong: it's not weapons but the morale which wins the wars. And if this is correct, perhaps one should take a second look at the budget of the Army, and maybe the budget could be cut. Colonel Adamson wheeled around to look at Mr. Wigner and said, "Well, as far as those $2,000 are concerned, you can have it." This is how the first money promise was made by the government. (*Docs. 66-70*)

I should mention that until the government showed interest (and the first interest it showed was the appointment of this committee) I was undecided whether this development ought to be carried by industry or whether it ought to be carried by the government. And so, just a week or two before the meeting in Washington, I had met

---

[14]See Szilard memorandum to Briggs, October 26, 1939, in *Scientific Papers*, pp. 204-206.

with the director of research of the Union Carbon and Carbide Company, W. F. Barrett. The appointment was made by Strauss, and there was some mix-up about it, because they expected Fermi and it was I who turned up.

I saw five people sitting around the table. I told them that the possibility of a chain reaction in uranium and graphite must be taken seriously; that at this point we could not say very much about this possibility; and that we could talk about it with much greater assurance if we first measured the absorption of neutrons in graphite. For this purpose we would need about $2,000 worth of graphite, and I wondered whether they might give us this amount of graphite on loan. The experiment would not damage the graphite and we could return it to them.

W. F. Barrett said, "You know, I'm a gambling man myself, but you are now asking me to gamble with the stockholders' money, and I'm not sure that I can do that. What would be the practical applications of such a chain reaction?" I said that I really cannot say what the practical applications would be at this point, that there is very little doubt in my mind that such a revolutionary phenomenon will find its practical applications ultimately, but it is too early to say that. We have first to see whether we can get it going, and under what conditions it can be set up.

After I left the meeting I had an uneasy feeling that I had not convinced anybody there. After all, I was a foreigner and my name was not so well-known; I was not well-known as a physicist, certainly not to these people. So I sat down and I wrote a letter to Mr. Barrett in which I invited him to lunch with Dr. Pegram (who was head of the department and dean of the graduate school) and Dr. Fermi (who after all was a Nobel Prize winner and his name was quite well-known) one day at his convenience the following week at Columbia University. (*Doc. 65*) He replied that the following week he would not be in town. He did not suggest an alternate date, and he wrote that they had decided that they would not be in a position to let us have any graphite except on a straight purchase basis. I remember that I was quite depressed by that letter. I showed it to Pegram, who thought that I was too easily discouraged. And maybe I was. [2]

Documents from April 1939 through December 1939.

After the breakdown of their attempt at secrecy, Szilard and his companions began to consider what other political steps they ought to take:

Document 49

[Wigner]
April 17, 1939

Dear Szilard!

Thank you for letting me have the news concerning the abandonment of any policy in the publication matter. I cannot help feeling, on the one hand, that this was, under the conditions, a wise decision as nothing really could be achieved in this matter, On the other hand I do feel, and I do feel it very strongly, that the U.S. Government should be advised of the situation. This is indicated, among *many other* reasons, by the necessity of preparing it to a possible sudden threat. Let me know, please, whether you have already taken steps in this direction and whether you intend to take some in the near future. . . .[1]

Hoping to hear from you very soon,

Sincerely,
Wigwam

Szilard tried without success to raise funds from the U.S. Army, the Union Minière du Haut-Katanga (owner of the Congo's uranium mines), and the U.S. Navy:

Document 50

Hotel King's Crown
New York City
May 21st, 1939

Dear Wigner:

It appears to me that it might be a good thing that you should establish contact with Kent[2] and visit him in Aberdeen some time. I would like to go along with you, but we are starting to-morrow on the experiment in which we use 500 pounds of uranium oxide,[3] and therefore I shall not be able to leave New York during the coming week. If necessary I could visit Mr. Kent at some later date, preferably after June 20th.

I am at a loss to find out what it is that Rabi told you, as the position is not changed in any way since I last saw you. There is in my opinion a 50% chance that a chain reaction might go with slow neutrons without separation of isotopes, though Fermi

---

[1] We omit a further three paragraphs dealing with getting a beryllium block from England.
[2] Probably Robert H. Kent, associate director of the ballistic research laboratories at the U.S. Army's Aberdeen Proving Ground, Maryland.
[3] See H. L. Anderson, E. Fermi, and L. Szilard, *Physical Review,* 56: 619–624 (1939), in *Scientific Papers,* pp. 160–162.

would perhaps put the chance smaller. Perhaps we shall know in about ten days from the experiment with the 500 pounds of uranium oxide.

I am rather anxious to see you as soon as possible, and if you do not come up to New York this week I would come up to Princeton one of these days, perhaps Wednesday in the late afternoon. If you go from Washington straight to Princeton we could have a talk there before you visit Aberdeen. I shall telephone to you at Princeton on Tuesday to find out whether you are back.

Please ask Teller to write down for you the calculations which he has made about the slow neutron reaction. We can then compare it with ours when we talk about these things at Princeton. . . .[4]

Please give my kind regards to Teller.

Yours,
Leo Szilard

Document 51

July 3rd, 1939

Mr. Lewis L. Strauss
52 William Street
New York City

Dear Mr. Strauss:

These are just a few lines to refresh your memory in case you find time to contact the Société Générale[5] or the Union Minière.

As you know Fermi and I made a number of experiments on uranium, some of these independently of each other, others jointly. All these experiments were carried out at the Physics Department of Columbia University with radium rented from the American agents of the Union Minière. In order to meet these and other expenses which would have strained the budget of the Department, other physicists and I formed an association called "Association for Scientific Collaboration" and collected some funds among ourselves.[6] I am writing to you in my capacity as one of the trustees of the Association rather than on behalf of the Physics Department, as I have not yet discussed the matter with the Head of the Department and have no authority to speak in the name of the Department.

A joint paper by Anderson, Fermi and myself, which has been recently completed and is not yet published, states that a nuclear chain reaction could be maintained under certain conditions in uranium, but expresses serious doubt whether such a chain reaction can be maintained in uranium oxide, or in uranium oxide mixed with water. It is my personal opinion that a chain reaction leading to the formation of practically unlimited amounts of radio-active material is an immediate possibility, though it requires careful control of the conditions under which the experiment is performed.

---
[4]We omit one paragraph speculating on ways to make a bomb.
[5]Parent organization of the Union Minière.
[6]The Association was incorporated in New York state. Its only funds were those Szilard borrowed or donated.

There is also a 50 to 50 chance that the matter may be of great importance from the point of view of national defense.

In the circumstances it seems to me that this work should be supported financially in a more efficient way than hitherto, and I wondered whether the Union Minière, as the company who would have the chief financial benefit of any use to which uranium is put, would be willing to support this work.

This support could take various forms: First of all, we would appreciate assurance that we can obtain at short notice 30 tons of uranium oxide on loan, of which we would probably need 5 tons in the next six weeks. Secondly it would be of great help if we need not pay rent for radium up to 2.5 gm, as long as we use it in the form in which the company has it in stock rather than in some other form. Thirdly, we would appreciate it if we could obtain refund of the rent which we paid between February and to date for radium loaned out of the stocks of the Radium Chemical Co., the American agent of the Radium Belge.

I am mentioning these details though I feel that a general agreement for co-operation along some broad lines would do much towards obtaining results quickly.

I personally have no doubt that if the Union Minière would prefer to loan the uranium to Columbia University rather than to a comparatively young "association," such formal difficulties that might exist could be overcome by general good will.

I would much appreciate if you could let me know by cable from England whether you were able to see any of the directors of the Union Minière by the end of July, because if you see for instance that you will not find time to deal with this matter I would like to attempt to contact the Union Minière at once through Professor Biot[7] who is at present in Belgium.

Forgive me please for troubling you with all this on the last day, but I assume that you might find time to read this letter on the boat.

Wishing you and Mrs. Strauss a pleasant stay in Europe.

I am, yours sincerely,

Leo Szilard

Document 52

[Gunn]
Naval Research Laboratory
Anacostia Station
Washington, D.C.
10 July 1939

Dr. Leo Szilard
Department of Physics
Columbia University
New York, N.Y.

Dear Dr. Szilard:

The matter which we discussed at the Princeton meeting of the Physical Society has

[7]Probably the Belgian physicist Maurice Anthony Biot.

been carefully considered. As I indicated to you at that time, it seems almost impossible, in light of the restrictions which are imposed on Government contracts for services, to carry through any sort of an agreement that would be really helpful to you. I regret this situation but see no escape. We are anxious, however, to cooperate with you in every respect and appreciate your assistance on this important problem.

Very truly yours,
Ross Gunn,
Technical Adviser

The following group of letters traces the events leading up to Einstein's letter to President Roosevelt:

Document 53

Hotel King's Crown
New York City
19 July 1939

Dear Professor [Einstein],

When I returned to New York after seeing you I found a message from Dr. Stolper, the former editor of the German Volkswirt. He reported to me that he had discussed our problems with Dr. Alexander Sachs, a vice-president of the Lehman Corporation, biologist and national economist, and that Dr. Sachs wanted to talk to me about this matter. I had turned to Dr. Stolper some time ago because I thought that we would have to draw on outside financial help for the large experiments, which exceeded the budget of the Physics Department. I now told Dr. Sachs everything in some detail, and Wigner's point of view that one should bring a large quantity of ore to America and alert the Belgian government to the situation, including our plan to write to the Belgian Ambassador and the State Department. Dr. Sachs took the position, and completely convinced me, that these were matters which first and foremost concerned the White House and that the best thing to do, also from the practical point of view, was to inform Roosevelt. He said that if we gave him [Sachs] a statement he would make sure it reached Roosevelt in person. Although I have seen Dr. Sachs once, and really was not able to form any judgment about him, I nevertheless think that it could not do any harm to try this way and I also think that in this regard he is in a position to fulfill his promise.

Then today I told Dr. Teller about this matter. Dr. Teller has been in the conspiracy with us from the beginning and is spending the summer here as a visiting professor in the Physics Department. (At other times he is in Washington.) Dr. Teller is also of the opinion that it would be preferable for us to take this approach. I was unable to reach Wigner, who is on his way to California and will not arrive there for a few days.

I have tried to draft a letter containing what I believe should be said, and am enclosing this draft herewith. I will then telephone you tomorrow to ask first of all whether, in principle, you agree with this procedure. If this is the case, perhaps you will be able to tell me over the telephone whether you would like to return the draft with your marginal comments by mail, or whether I should come out to discuss the whole thing once more with you. If you wish me to come out I would like, if it is all right with you, to ask Teller to take me, not only because I believe his advice is valuable but also because I think you might enjoy getting to know him. He is particularly nice.[8]

Yours very sincerely,
Leo Szilard

[German original of the preceding letter]

19 July 1939

Lieber Herr Professor,

Als ich von Ihnen nach New York zurückkehrte, fand ich eine Nachricht von Dr. Stolper, der frühere Redakteur des Deutschen Volkswirt, vor, der mir mitteilte, dass er über unsere Problems mit Dr. Alexander Sachs, ein Vice-President der Lehman Corporation, Biologe und Nationalökonome, gesprachen hat und dass Dr. Sachs mich in dieser Angelegenheit sprechen wollte. Ich hatte mich vor einiger Zeit an Dr. Stolper gewandt weil ich glaubte, dass wir für die grossen Versuche, die über das Budget des Physics Department hinausgehen, finanzielle Hilfe von aussen werden heranziehen müssen. Ich habe nun Dr. Sachs ziemlich ausführlich alles erzählt, auch Wigners Standpunkt, dass man eine grosse Menge Erzes nach Amerika bringen und die belgische Regierung auf die Situation aufmerksam machen soll, einschliesslich unserem Plan an den belgischen Ambassador und an das State-Department zu schreiben. Dr. Sachs vertrat den Standpunkt und er hat mich vollständig überzeugt, dass diese Angelegenheit in erster Linie das Weisse Haus angeht und es auch vom praktischen Standpunkt das beste ist Roosevelt zu informieren. Er sagte, dass wenn wir ihm ein statement geben, er dafür sorgen wird, dass dieses in die Hände von Roosevelt gelangt. Obwohl ich Dr. Sachs nur einmal im Leben gesehen habe und mir eigentlich kein Urteil über ihn bilden konnte, so glaube ich doch, dass es nicht schaden könnte wenn wir diesen Weg versuchen und ich glaube auch, dass er in der Lage ist, in dieser Beziehung das zu halten was er versprochen hat.

Ich habe dann noch heute mir Dr. Teller, der von Anfang an mit bei der Verschwörung dabei war und der jetzt im Sommer hier als Gast-Professor im Physics Department ist (sonst ist er in Washington) von der Sache erzählt und auch er war der Ansicht, dass wir lieber diesem Weg beschreiten sollen. Wigner konnte ich nicht mehr erreichen, er ist unterwegs nach Californien und wird erst in einigen Tagen dort ankommen.

Ich habe versucht einen Brief aufzusetzen, der das enthält was ich glaube, dass gesagt werden sollte, und schicke Ihnen diesen Entwurf in der Beilage ein. Ich werde Sie morgen telephonisch anrufen um Sie zunächst mal zu fragen, ob Sie überhaupt

---

[8]Our translation. Enclosed in this letter from Szilard was a double spaced 4 1/2-page English draft of a letter addressed to the President of the United States, Washington, D.C., prepared for Einstein's signature.

mit der ganzen Prozedur im Prinzip einverstanden sind. Falls dies der Fall ist, werden Sie mir dann im Telephon vielleicht sagen können, ob Sie mir den Entwurf mit Ihren Randbemerkungen per Post zurückschicken wollen oder ob ich hinausfahren soll, um mit Ihnen noch einmal über die ganze Sache zu sprechen. Falls ich hinauskommen soll, würde ich gerne, wenn es Ihnen recht ist, Teller bitten, mich hinauszufahren, und zwar sowohl weil ich glaube, dass sein Rat von Wert ist, wie auch weil ich glaube, dass es Ihnen Freude machen würde ihn kennen zu lernen. Er ist besonders nett.

Ihr sehr ergebener
Leo Szilard

Document 54

Hotel King's Crown
New York City
August 2, 1939

Dear Professor [Einstein]:

It would be very good if we could at long last decide upon whom we should try to get as middleman. Dr. Sachs from the Lehman Corporation, with whom I have talked again in the meantime, suggested Bernard Baruch[9] or K. T. Compton[10] on a trial basis, but they do not seem very suitable to me nor to Teller. On the other hand Sachs thinks that Lindbergh[11] is a good choice so at the moment Lindbergh is the "favorite." We shall have to rack our brains some more.

In the meantime I would like to try to talk with Lindbergh about the matter to see how he feels about it. I met him about 7 years ago, and liked him, but I assume he will have forgotten me by now. If you could perhaps re-introduce me I would ask you to send me a letter of introduction (addressed to Col. Charles Lindbergh) to New York. I could then enclose it with a letter which I am writing to Lindbergh myself.

I am enclosing the German text which we drafted together in Peconic and the English translation. Also enclosed is a somewhat longer and more extensive version which I drafted after my last discussion with Sachs. The first version has the advantage of brevity, but the second contains everything necessary to give the president a clear picture of what duties would have to be carried out by the person he would delegate. I do not know which of the two will seem more appropriate to you and am therefore sending you both.

If by any chance neither version meets with your approval, you could perhaps send me an edited German copy which I could then return to you translated into English.

Don't you think that it might be good if I were to speak also to Gano Dunn?[12] To be sure I have not seen him for 7 years and it is said that he had lost a large part of

---

[9]Bernard M. Baruch, financier.
[10]Karl T. Compton, president of the Massachusetts Institute of Technology.
[11]The famous aviator Charles Lindbergh, who returned to the U.S. in 1939. See *Doc. 59*.
[12]Electrical engineer.

his fortune and is no longer president of the G. J. White Corporation. But he knows many people and his advice could therefore be very useful. I think I will write to him; I am sure I can dig up his current address somehow or other.

Also enclosed is the manuscript of which we spoke. On this subject there are some rather interesting questions which it might be useful to discuss at some time or other. If you liked Teller, I would like to come out to your place some time or other with him. Though it has just occurred to me that he will only be in New York until the 13th.

Yours respectfully,
Leo Szilard[13]

[German original of the preceding letter]

Hotel King's Crown
New York City
den 2. August 1939

Sehr geehrter Herr Professor:

Es waere sehr gut, wenn wir uns allmaehlich darueber klar werden koennten, wessen Ernennung als Mittelsperson man anstreben sollte. Dr. Sachs von der Lehman Corporation, mit dem ich inzwischen wieder gesprochen habe, hat Bernhard Baruch oder K. T. Compton versuchsweise vorgeschlagen, doch scheinen sowohl Teller wie auch mir diese Personen nicht sehr geeignet, Dagegen haelt Sachs die Wahl von Lindbergh fuer gut, sodass im Augenblick Lindbergh der "Favorit" ist. Wir werden uns noch weiter den Kopf zerbrechen muessen.

Inzwischen wurde ich jedenfalls gerne versuchen, mit Lindbergh ueber die Sache zu sprechen, um zu sehen, wie er sich dazu stellt. Ich habe ihn vor etwa 7 Jahren getroffen, er hat mir gut gefallen, aber ich nehme an, dass er mich inzwischen vergessen hat. Wenn Sie mich etwa re-introducen koennen, so wuerde ich Sie bitten, mir ein Einfuehrungsschreiben (adressiert an Col. Charles Lindbergh) nach New York zu schicken, welches ich einem Brief, den ich selber an Lindbergh schreibe, beilegen wuerde.

In der Anlage schicke ich Ihnen den deutschen Text, den wir zusammen in Peconic aufgesetzt haben, und die englische Uebersetzung dazu. Ebenfalls in der Beilage schicke ich eine etwas laengere und ueber den deutschen Text hinausgehende Fassung, die ich nach der letzten Besprechung mit Sachs aufgesetzt habe. Die erste Fassung hat den Vorzug der Kuerze, dagegen enthaelt die zweite all das, was noetig ist, um dem Praesidenten ein klares Bild zu geben, welche Aufgaben die von ihm zu bestimmende Vertrauensperson zu erfuellen haette. Ich weiss nicht, welche von den beiden Fassungen Ihnen als richtiger erscheinen wird und schicke Ihnen daher beide Fassungen zu.

Falls Ihnen keine der beiden gefaellt, koennten Sie mir vielleicht einen veraenderten deutschen Text zuschicken, den ich Ihnen dann ins Englische uebertragen zuruecksenden wuerde.

[13]Our translation.

Glauben Sie nicht, dass es vielleicht gut waere, wenn ich auch Gano Dunn sprechen wuerde? Ich habe ihn allerdings seit 7 Jahren nicht gesehen und es heisst, dass er einen grossen Teil seines Vermoegens verloren hat und nicht mehr Praesident der G. J. White Corporation ist. Er kennt viele Leute, und sein Rat koennte daher sehr nuetzlich sein. Ich denke, ich werde an ihn schreiben, seine jetzige Adresse wird man schon irgendwie auftreiben koennen.

In der Anlage schicke ich Ihnen auch das Manuskript, von dem die Rede war. Es gibt in diesem Zusammenhang ganz interessante Fragen, ueber die es sich vielleicht gelegentlich zu sprechen lohnte, und wenn Ihnen Teller gefallen hat, so wuerde ich gern gelegentlich wieder einmal mit ihm zu Ihnen hinausfahren. Er ist allerdings, wie mir eben einfaellt, nur noch bis zum 13. in New York.

Ihr sehr ergebener
Leo Szilard

Enclosed in the preceding letter were:

1. A typed transcription of the German text Szilard wrote down in Peconic on the second visit ("Die in Peconic aufgesetzte Formulierung").

2. Its translation into an English letter to the President for Einstein's signature.[14]

3. The longer version of the same letter given in the following.[15]

4. Preprint dated June 22, 1939 of H. L. Anderson, E. Fermi, and Leo Szilard, "Neutron Production and Absorption in Uranium," Physical Review, *56: 284–286 (1939), in* Scientific Papers, *pp. 163–168.*

Document 55

Albert Einstein
Old Grove Rd.
Nassau Point
Peconic, Long Island
August 2nd, 1939

F. D. Roosevelt
President of the United States
White House
Washington, D.C.

Sir:

Some recent work by E. Fermi and L. Szilard, which has been communicated to me

---

[14]This shorter version of the letter signed by A. Einstein is in the Szilard files, with a pencilled mark in Szilard's handwriting—"original, not sent." The chief differences between the two versions, aside from minor rearrangements, are the addition, in the longer version, of the section from "that the element uranium . . ." to ". . . Fermi and Szilard in America"; of the clause "and large quantities of new radium-like elements"; of the sentence beginning "Now it appears almost certain. . . ."; and of the section "His task might comprise the following: a) . . . b) . . . ."

[15]A copy of this longer version is in the Szilard files. It was also printed in *Einstein on Peace* (New York: Schocken Books, 1968), pp. 294–296, and in *Scientific Papers*, pp. 199–200.

Chapter III: Documents from April 1939 through December 1939

in manuscript, leads me to expect that the element uranium may be turned into a new and important source of energy in the immediate future. Certain aspects of the situation which has arisen seem to call for watchfulness and, if necessary, quick action on the part of the Administration. I believe therefore that it is my duty to bring to your attention the following facts and recommendations:

In the course of the last four months it has been made probable—through the work of Joliot in France as well as Fermi and Szilard in America—that it may become possible to set up a nuclear chain reaction in a large mass of uranium by which vast amounts of power and large quantities of new radium-like elements would be generated. Now it appears almost certain that this could be achieved in the immediate future.

This new phenomenon would also lead to the construction of bombs, and it is conceivable—though much less certain—that extremely powerful bombs of a new type may thus be constructed. A single bomb of this type, carried by boat and exploded in a port, might very well destroy the whole port together with some of the surrounding territory. However, such bombs might very well prove to be too heavy for transportation by air.[16]

The United States has only very poor ores of uranium in moderate quantities. There is some good ore in Canada and the former Czechoslovakia, while the most important source of uranium is the Belgian Congo.

In view of this situation you may think it desirable to have some permanent contact maintained between the Administration and the group of physicists working on chain reactions in America. One possible way of achieving this might be for you to entrust with this task a person who has your confidence and who could perhaps serve in an inofficial capacity. His task might comprise the following:

a) to approach Government Departments, keep them informed of the further development, and put forward recommendations for Government action, giving particular attention to the problem of securing a supply of uranium ore for the United States,

b) to speed up the experimental work, which is at present being carried on within the limits of the budgets of University laboratories, by providing funds, if such funds be required, through his contacts with private persons who are willing to make contributions for this cause, and perhaps also by obtaining the co-operation of industrial laboratories which have the necessary equipment.

I understand that Germany has actually stopped the sale of uranium from the Czechoslovakian mines which she has taken over. That she should have taken such early action might perhaps be understood on the ground that the son of the German

---

[16]Szilard, and all other physicists who considered the problem in 1939, thought of a bomb as a slow-neutron reaction, a sort of giant reactor—which would not in fact be very explosive. The fast-neutron bomb, using pure uranium-235 or plutonium, was not yet imagined.

Under-Secretary of State, von Weizsäcker,[17] is attached to the Kaiser-Wilhelm-Institut in Berlin where some of the American work on uranium is now being repeated.[18]

Yours very truly,
A. Einstein

Einstein answered Szilard with a handwritten note in German saying that while he had signed both letters he would give preference to the more detailed version. The requested introduction to Lindbergh was also enclosed. Einstein added that in any case he hoped that Szilard and his colleagues would finally overcome their inner resistances, because it was always questionable if one wanted to do something all too cleverly. Szilard's reply to this letter is the following document.

Document 56

August 9, 1939

Dear Professor [Einstein]:

I am just writing to acknowledge receipt of your letter with the enclosures. We will try to follow your advice and as far as possible overcome our inner resistances which, admittedly, exist. Incidentally, we are surely not trying to be too clever and will be quite satisfied if we don't do things too foolishly. I will keep you informed on how things progress.[19]

Yours,
Szilard

[German original of the preceding letter]

den 9. August 1939

Sehr geehrter Herr Professor!

Ich schreibe Ihnen nur, um zu bestaetigen, dass Ihr Brief mit den Beilagen uns erreicht hat. Wir werden versuchen, Ihre Mahnung zu beherzigen und so gut es geht unsere

---

[17]The son, Carl F. von Weizsäcker, did get involved in chain reaction work.

[18]Roosevelt replied October 19, 1939:

My dear Professor:
I want to thank you for your recent letter and the most interesting and important enclosure.
I found this data of such import that I have convened a board consisting of the head of the Bureau of Standards and chosen representatives of the Army and Navy to thoroughly investigate the possibilities of your suggestion regarding the element of uranium.
I am glad to say that Dr. Sachs will cooperate and work with this Committee and I feel this is the most practical and effective method of dealing with the subject.
Please accept my sincere thanks.

Very sincerely yours,
Franklin D. Roosevelt

[19]Our translation.

inneren Widerstaende, die zugegebenermassen vorhanden sind, zu ueberwinden. Uebrigens wollen wir es gar nicht gar zu gescheit machen und sind schon zufrieden, wenn wir es nicht gar zu dumm machen. Ich werde Ihnen laufend berichten, wie es weitergeht.

Ihr
Szilard

Document 57

Hotel King's Crown
New York City
August 9th, 1939

Dear Wigner:

As Teller has already told you it has been decided, following up your initiative, to withhold the letter to the Belgian Ambassador and to substitute for the letter to the State Department a rather different letter to the President. This letter, which is written by E. and has been sent to me to be forwarded, will not be sent by post but handed over to a vice-president of the Lehman Corporation who promised that he will see to it that it gets attention. The only suggestion made in that letter is that the President appoint a person who could serve—perhaps in an inofficial capacity—as a Permanent link between the Administration and the physicists and fulfill a double function, i.e. make recommendations to Government departments and see to it that private funds are provided for accelerating the experimental development.

It seems to me important that you should not talk about this to *any* third person. I shall let you know if anything develops out of this attempt. Perhaps you could drop me a line, letting me know what your present address is.

Yours,
Leo Szilard

Document 58

[Letter of transmittal, Szilard to Dr. Alexander Sachs]

August 15, 1939

Dear Dr. Sachs:

Enclosed I am sending you a letter from Prof. Albert Einstein, which is addressed to President Roosevelt and which he sent to me with the request of forwarding it through such channels as might appear appropriate. If you see your way to bring this letter to the attention of the President, I am certain Prof. Einstein would appreciate your doing so; otherwise would you be good enough to return the letter to me?

If a man, having courage and imagination, could be found and if such a man were put—in accordance with Dr. Einstein's suggestion—in the position to act with some measure of authority in this matter, this would certainly be an important step forward. In order that you may be able to see of what assistance such a man could be in our work, allow me please to give you a short account of the past history of the case.

In January this year, when I realized that there was a remote possibility of setting up a chain reaction in a large mass of uranium, I communicated with Prof. E. P. Wigner of Princeton University and Prof. E. Teller of George Washington University, Washington, D.C., and the three of us remained in constant consultation ever since. First of all it appeared necessary to perform certain fundamental experiments for which the use of about one gram of radium was required. Since at that time we had no certainty and had to act on a remote possibility, we could hardly hope to succeed in persuading a university laboratory to take charge of these experiments, or even to acquire the radium needed. Attempts to obtain the necessary funds from other sources appeared to be equally hopeless. In these circumstances a few of us physicists formed an association, called "Association for Scientific Collaboration," collected some funds among ourselves, rented about one gram of radium, and I arranged with the Physics Department of Columbia University for their permission to carry out the proposed experiments at Columbia. These experiments led early in March to rather striking results.

At about the same time Prof. E. Fermi, also at Columbia, made experiments of his own, independently of ours, and came to identical conclusions.

A close collaboration arose out of this coincidence, and recently Dr. Fermi and I jointly performed experiments which make it appear probable that a chain reaction in uranium can be achieved in the immediate future.

The path along which we have to move is now clearly defined, but it takes some courage to embark on the journey. The experiments will be costly since we will now have to work with tons of material rather than—as hitherto—with kilograms. Two or possibly three different alternatives will have to be tried; failures, set-backs and some unavoidable danger to human life will have to be faced. We have so far made use of the Association for Scientific Collaboration to overcome the difficulty of persuading other organisations to take financial risks, and also to overcome the general reluctance to take action on the basis of probabilities in the absence of certainty. Now, in the face of greater certainty, but also greater risks, it will become necessary either to strengthen this association both morally and financially, or to find new ways which would serve the same purpose. We have to approach as quickly as possible public-spirited private persons and try to enlist their financial cooperation, or, failing in this, we would have to try to enlist the collaboration of the leading firms of the electrical or chemical industry.

Other aspects of the situation have to be kept in mind. Dr. Wigner is taking the stand that it is our duty to enlist the co-operation of the Administration. A few weeks ago he came to New York in order to discuss this point with Dr. Teller and me, and on his initiative conversations took place between Dr. Einstein and the three of us. This led to Dr. Einstein's decision to write to the President.

I am a enclosing memorandum which will give you some of the views and opinions which were expressed in these conversations.

I wish to make it clear that, in approaching you, I am acting in the capacity of a trustee of the Association for Scientific Collaboration, and that I have no authority to speak in the name of the Physics Department of Columbia University, of which I am a guest.

Yours sincerely,
Leo Szilard

Document 59

August 16th, 1939

Colonel Charles Lindbergh
Washington, D.C.

Dear Colonel Lindbergh:[20]

I had the pleasure of meeting you at lunch about seven years ago at the Rockefeller Institute, but I assume that you do not remember me, and I am therefore enclosing an introduction by Prof. Albert Einstein.[21]

Experiments on uranium, which we have recently performed, lead to the conclusion that a nuclear chain reaction could be maintained under certain specific conditions in a large mass of uranium. Large quantities of energy would be liberated, and great amounts of new radio-active elements would be generated in such a chain reaction. The experiments have reached a point where it becomes necessary to work with tons of material rather than with the kilograms hitherto used. The path along which we have to walk is clearly defined, but it requires some courage to embark on the journey.

At recent discussions, which took place at the house of Prof. Einstein and in which Prof. E. P. Wigner of Princeton University and Professor E. Teller of George Washington University participated, there was a feeling that this matter has taken proportions which make it somewhat difficult for us to cope with the problems connected with it. It has therefore been decided to make an attempt to inform the Administration of the United States of the situation which has arisen. This will probably be done in the near future. In this connection it appeared that it might be of value to discuss the problem with you, and we should certainly appreciate having your advice or help in this matter. You might perhaps find some difficulty in getting acquainted with the details, but it will perhaps interest you to obtain a general idea of the outlines and to learn of the new facts which came to light during the last few months and which seem to lead to such a dramatic development.

Should you happen to visit New York or be near New York, would you be kind

---
[20]We do not know whether this letter was actually sent.
[21]We have not found the introduction by Professor Einstein in our files and this might mean that it was sent along with the letter.

enough to let me know if you care to see me? I get there about once a month, and, if you are not going to visit New York in the near future, perhaps you might find it convenient to see me in Washington.

Yours very truly,
Leo Szilard

Meanwhile Szilard continued to seek support from individuals and industry:

Document 60

Hotel King's Crown
New York City
September 27th, 1939

Dear Professor [Einstein]:

Enclosed is the talk by Lindbergh which you have perhaps not read. I am afraid he is in fact not our man. Besides, the discussion about the neutrality law is on a pitiful level. At that one becomes kindly disposed towards Lindbergh for he at least emits human sounds.

As I understand it your letter to the President has been in Washington for some time. I shall see Dr. Sachs (Lehman Corporation) on Friday and shall perhaps hear whether anything has ensued.

Since we must be prepared for the fact that Belgium will be overrun one of these days, I want to attempt now to see that at least 50 tons of uranium oxide is purchased. It can always be sold on the market—perhaps even with a profit—if and when the material is no longer needed. Whether I will be able to persuade a government agency to take such a step I do not know of course. Perhaps one would have more luck with a smart speculator.[22]

With kind regards,

Yours very sincerely,
Leo Szilard

[German original of the preceding letter]

Hotel King's Crown
420 West 116th Street
New York City
September 27th, 1939

Lieber Herr Professor!

Ich schicke Ihnen in der Anlage die Rede von Lindbergh, die Sie vielleicht nicht gelesen haben. Ich fuerchte, er ist in der Tat nicht unser Mann. Im uebrigen steht die

---

[22]Our translation.

Diskussion ueber das Neutralitaetsgesetz auf einem erbaermlichen Niveau. Man wird dabei Lindbergh gegenueber noch ganz milde gestimmt, denn er gibt wenigstens menschliche Toene von sich.

Soviel ich verstehe, ist Ihr Brief an den Praesidenten schon seit einiger Zeit in Washington. Ich sehe Dr. Sachs (Lehman Corporation) am Freitag und werde vielleicht hoeren, ob irgendetwas erfolgt ist.

Da wir darauf gefasst sein muessen, dass Belgien an einem dieser Tage ueberrannt wird, will ich versuchen jetzt durchzusetzen, dass wenigstens 50 Tonnen Uranoxide gekauft werden, die man ja spaeter, wenn das Material nicht mehr gebraucht wird, jederzeit auf dem Markt—vielleicht sogar mit Gewinn—verkaufen kann. Ob es gelingen wird, eine Regierungsstelle zu einem solchen Schritt zu bewegen, weiss ich natuerlich nicht. Vielleicht wuerde man es mit einem klugen Spekulanten leichter haben.

Mit freundlichen Gruessen

Ihr sehr ergebener
Leo Szilard

Document 61

King's Crown Hotel
New York City
3 October 1939

Dear Professor [Einstein]:

Last week Wigner and I visited Dr. Sachs, who admitted that he is still holding your letter. He says he has spoken repeatedly with Roosevelt's secretary and has the impression that Roosevelt is so overburdened that it would be wiser to see him at a later date. He intends to go to Washington this week.

There is a distinct possibility that Sachs will be of no use to us. If this is the case, we must put the matter in someone else's hands. Wigner and I have decided to accord Sachs ten days' grace. Then I will write you again to let you know how matters stand.[23]

With friendly greetings,

Yours very sincerely,
Leo Szilard

[German original of the preceding letter]

den 3. 10. 1939

Lieber Herr Professor!

Wigner und ich haben vorige Woche Dr. Sachs besucht, der uns gestanden hat, dass er

---
[23]Our translation.

immer noch auf Ihrem Brief sitzt. Er sagte, er haette wiederholt mit dem Sekretaer von Roosevelt telephoniert und den Eindruck gewonnen, dass Roosevelt so ueberlastet ist, dass es klueger waere, ihn spaeter zu sehen. Er hatte vor, diese Woche nach Washington zu fahren.

Es ist durchaus moeglich, dass Sachs unbrauchbar ist. Gegebenenfalls muessen wir die Sache in die Hand eines Anderen legen. Ich habe mit Wigner verabredet, dass wir Sachs noch eine Frist von 10 Tagen lassen. Dann werde ich Ihnen wieder schreiben, wie die Sache steht.

Mit freundlichen Gruessen

Ihr sehr ergebener
Leo Szilard

Document 62

September 13th, 1939

Mr. Gano Dunn
80 Broad Street
New York City

Dear Mr. Dunn:

I wonder if you still remember me. About six years ago you were kind enough frequently to help me with your advice. This was in connection with a new type of household refrigerator which Professor Einstein and I jointly devised and for which we tried to find interest in this country.

Today I am writing to you about a very different matter: Experiments on uranium, with which I have been connected during the last six months, lead to the conclusion that a nuclear chain reaction could be maintained under certain specific conditions in a large mass of uranium. Large quantities of energy would be liberated and great amounts of new radioactive elements would be generated in such a chain reaction. The experiments have now reached a point which makes it necessary to proceed on an almost industrial scale, working with tons of the material rather than the minute quantities hitherto used.

It would seem that this is a matter which requires the cooperation of a number of people and organizations, and it would hardly seem proper to put the matter in the hands of one single industrial corporation and to look upon it primarily from a business point of view.

I would very much appreciate having an opportunity to discuss this matter with you and to have your advice.

Yours very sincerely,
Leo Szilard

P. S. Enclosed you will find a reprint of the first of a series of papers dealing with this subject.

Document 63

Eugene P. Wigner
120 Prospect Avenue
Princeton, New Jersey
1939 september 26

Dear Szilard:

I am afraid that there is some misunderstanding between us. When you were in Princeton a few days ago, I got the impression that the work at Columbia University is near to a standstill and that the next step should consist in large scale experiments for which both funds and proper equipment is lacking at Columbia. I understood that you are considering therefore to shift the place of further work to some other location and that it is in this connection that you wanted to talk to [K. T.] Compton. It was under this impression that I wrote the letter the copy of which you will find enclosed.

After having written this letter, I was considering the matter again. I found that the procedure is not a practicable one. Compton would be naturally very careful and even if it could be made clear to him that it is desirable to continue the experiments at a new location, it is quite doubtful whether or not he would be able to entrust them (and the management of funds) to you. And, of course we cannot blame him for this, as he is not well acquainted with you.

I am somewhat doubtful whether it would be at all possible to induce Compton to find money for experiments conducted by you as he would naturally (and quite incorrectly) feel that the experiments, if your cooperation with Fermi is broken up, may better be conducted by a third person. But if it is possible to induce Compton to this, the proper procedure certainly would be that a third person (perhaps I) should see him and present to him at this occasion a memorandum written by you. As I said, however, I am even in this case quite doubtful that this is a matter worth trying, it looks so difficult.

I should mention, perhaps, that I was also considering to secure Compton's help for experiments to be done at Columbia. I felt, however, that such a step should be undertaken by Fermi and if he is unwilling to do this for some reason (which would be quite regrettable) you should have at least a letter from him endorsing you. Otherwise, everybody would ask the question as to the reason for your acting instead of Fermi.

When I just talked to you over the telphone, I understood that all you want to propose to Compton is the purchase of 50 tons of Uranium Oxide. I think that even as far as this is concerned, the prospect for a success would be greater if Fermi did act. However, this could be done, perhaps by you also. I am under the impression, however, that it would be better if you wrote a short memorandum on this question which you could send to Compton (or I could send to him in your name). I believe that this would make a better impression on him and further possible future relations with him to a greater extent. It would be also easier for him to follow up this particular matter as he would be able to consult the memorandum whenever necessary. From a conversation with you, if he had no opportunity to acquaint himself with the matter before, he would probably obtain only a somewhat confused picture which is undesirable.

I hope that I succeeded in making myself clear and that you will agree with me and see my point.

Sincerely,
Wigner

[Marginal note by Wigner translated from Hungarian]

Please excuse the "hard" tone of this letter, but I feel that in this manner I can make it clearer what I think. Farkas, who is here, also believes that we'll have a better situation if Compton has something in writing from you before you talk to him.
Regards, Wigwam.

The Einstein letter was finally delivered to the President, under circumstances documented in the following:

Document 64

[Sachs]
October 11, 1939

Dear Mr. President

With approaching fulfillment of your plans in connection with revision of the Neutrality Act, I trust that you may now be able to accord me the opportunity to present a communication from Dr. Albert Einstein to you and other relevant material bearing on experimental work by physicists with far-reaching significance for National Defense.

Briefly, the experimentation that has been going on for half a dozen years on atomic disintegration has culminated this year (a) in the discovery by Dr. Leo Szilard and Professor Fermi that the element, uranium, could be split by neutrons[24] and (b) in the opening up of the probability of chain reactions,—that is, that in this nuclear process uranium itself may emit neutrons. This new development in physics holds out the following prospects:

---

[24]This discovery was in fact made by Hahn and Strassmann.

1. The creation of a new source of energy which might be utilized for purposes of power production;

2. The liberation from such chain reaction of new radio-active elements, so that tons rather than grams of radium could be made available in the medical field;

3. The construction, as an eventual probability, of bombs of hitherto unenvisaged potency and scope: as Dr. Einstein observes, in the letter which I will leave with you, "a single bomb of this type carried by boat and exploded in a port might well destroy the whole port together with some of the surrounding territory! . . ."[25]

In view of the danger of German invasion of Belgium, it becomes urgent to make arrangements—preferably through diplomatic channels—with the Union Minière du Haut-Katanga, whose head office is at Brussels, to make available abundant supplies of uranium to the United States. In addition, it is necessary to enlarge and accelerate the experimental work, which can no longer be carried out within the limited budgets of the departments of theoretical physics in our universities. It is believed that public-spirited executives in our leading chemical and electrical companies could be persuaded to make available certain amounts of uranium oxide and quantities of graphite, and to bear the considerable expense of the newer phases of the experimentation. An alternative plan would be the enlistment of one of the foundations to supply the necessary materials and funds. For either plan and for all the purposes, it would seem advisable to adopt the suggestion of Dr. Einstein that you designate an individual and a committee to serve as a liaison between the scientists and the Executive Departments.

In the light of the foregoing, I desire to be able to convey in person, in behalf of these refugee scholars, a sense of their eagerness to serve the nation that has afforded them hospitality, and to present Dr. Einstein's letter together with a memorandum[26] which Dr. Szilard prepared after some discussion with me and copies of some of the articles that have appeared in scientific journals. In addition, I would request in their behalf a conference with you in order to lay down the lines of policy with respect to the Belgian source of supply and to arrange for a continuous liaison with the Administration and the Army and Navy Departments, as well as to solve the immediate problems of necessary materials and funds.

With high regard,

Yours sincerely,
Alexander Sachs

The President
The White House
Washington, D.C.

While preparing to meet with the Briggs Committee, which was set up in response to Einstein's letter, Szilard continued to seek private support for the work:

[25] We omit two paragraphs.
[26] This memorandum (printed in *Scientific Papers*, pp. 201–203) went into problems of obtaining uranium and funds, and the possibility of a bomb from a fast neutron chain reaction, in somewhat more detail than the Einstein letter.

Document 65

Hotel King's Crown
New York City
October 18, 1939

Mr. William F. Barrett, Vice-President
Union Carbide and Carbon Corp.
30 East 42nd Street
New York City

Dear Mr. Barrett:

Enclosed I am sending you a short memorandum referring to the conversation which we had on Monday this week. We have calculated how much graphite we would need for the preliminary experiment, which we propose to start immediately, and find that we could probably manage with 4 metric tons. As you will see from the enclosed copy of a letter of the National Carbon Co. the price quoted for this amount of graphite is about $3500. As soon as you let us know whether we can have this amount of graphite we would take all the necessary steps for preparing this experiment.

I have telephoned to your office today and left a message with Mr. Mills in order to ask you whether you would care to meet Professor Pegram, who is in charge of the Physics Department at Columbia University, and Professor Enrico Fermi one of these days for lunch at the Faculty Club. If you let me know what days would be convenient to you I would find out when the others are free and communicate with you.

On Friday I have to leave for Washington where a meeting has now been fixed for Saturday, but I hope to be back by Tuesday at the latest.

Yours very sincerely,
Leo Szilard

[Enclosure in preceding letter]

October 18, 1939

MEMORANDUM

Recent results concerning the possibility of setting up a nuclear chain reaction in uranium make it appear desirable that we should establish some sort of cooperation with the Union Minière. Just what form of cooperation would be most appropriate has not been decided as yet. I have seen Mr. Jean E. V. Cattier, whose father is President of the Union Minière, and arranged with him to meet in the near future the Managing Director, Mr. [Edgar] Sengier, who is now here on a visit.

It appeared desirable that the firms who use American uranium ores should be contacted before any definite arrangements are made with the Union Minière, especially since, in an emergency, the United States might be cut off from Canadian and Belgian supplies. I was advised that most of the carnotites containing uranium are mined by the Vanadium Corporation of America, which is a subsidiary of the Union

Carbide and Carbon Corporation. This was primarily the reason for my contacting the Carbide and Carbon Corporation.

A second reason for my contacting this corporation was the fact that an experiment is being considered for which about 50 tons of graphite might be required. This experiment also requires a large quantity of uranium oxide. It is assumed that it will be possible to obtain the uranium oxide required for this experiment as a loan from the Union Minière. While it is impossible to say with certainty that this experiment will lead to a large scale liberation of nuclear energy, there is a good chance that this will be the case. Obviously, the success of the experiment would lead to a great increase in the value of uranium and would thereby directly benefit those who control the supply of uranium ores. The cooperation of the Union Minère and of the Union Carbide and Carbon Corporation would appear to be justified on this ground alone, even without emphasizing the possible importance of these experiments from the point of view of national defense.

Fifty metric tons of graphite, which might possibly be required for the experiment which is envisaged, represent a value of about $16,000.00 if ordinary graphite is used, which rates at a price of 15 cents per pound. However, it seems that a special brand of [very pure] graphite has to be used, for which the National Carbon Company quotes a price of 35 cents per pound; the value involved would then be about $40,000.00. A letter of the National Carbon Company containing a price quotation is enclosed.

It seems impossible to foresee with certainty the outcome of the proposed large scale experiment, and it appears that we simply will have to have the courage to embark on it on a fifty-fifty chance for success and failure.

The estimate of the chances of this experiment might be slightly improved by investigating the properties of graphite in a separate experiment which we propose to start as soon as possible. This requires 4 metric tons of graphite of the grade specified in the enclosed letter of the National Carbon Company, and on the basis of the quotation contained in this letter this amount would represent a value of about $3500.00. The Physics Department at Columbia University has at present no funds available for the purpose of this experiment, but it is believed that, if the National Carbon Company would consent to supply the graphite material, the other facilities required could be obtained from the Rockefeller Foundation or some other Foundation. An early decision on this point would be appreciated.

Leo Szilard

In the following Szilard comments on the first meeting of the Briggs Committee:

Document 66

Hotel King's Crown
New York City
October 17, 1939

Dear Professor [Einstein]:

I trust that Wigner has informed you by telephone that Dr. Sachs (of Lehman Corporation) was in Washington last week, handed your letter to the President personally and that the President read this letter attentively. Dr. Sachs spoke with his secretary last Saturday, she spoke on the telephone with me, I spoke on the telephone with Wigner and Wigner telephoned you. Since this form of transmitting news is perhaps too indirect, I wish to report to you today by letter, having spoken with Sachs yesterday.

In your letter to Roosevelt you suggested that a connecting link be created which would keep up the contact between the Administration and the physicists. Roosevelt then asked Sachs which method he would suggest for carrying out your proposal. Dr. Sachs suggested the nomination of a committee of not more than three people for this purpose.

Roosevelt accepted this suggestion and nominated a committee composed of Briggs, Head of the Bureau of Standards in Washington, a Colonel who no doubt represents the Army, and a commander who, I presume, represents the Navy. Briggs has written to Sachs and asked him to come to the first meeting of this committee (which is now set for Saturday morning) in Washington, and to bring Wigner and me with him so that someone will be there who can give information about technical details. Since it would be difficult for Wigner and me to travel to Washington too frequently, Sachs, at my request, called Briggs and arranged with him that Dr. Teller, who is always in Washington and whom Briggs of course knows, will be included in this conference. In this way we shall probably be able through Teller to keep contact with Briggs in a workable fashion.

If my somewhat stubborn cold allows, I shall come to Princeton before my trip to Washington, and I hope that at that time, if your schedule permits, we can discuss all the pending questions.[27]

With kind regards,

Yours very sincerely,
Leo Szilard

[German original of the preceding letter]

Hotel King's Crown
420 West 116th Street
New York City
den 17. Oktober 1939

Lieber Herr Professor!

Wigner hat Ihnen wohl telephonisch berichtet, dass Dr. Sachs (von Lehman Corporation) vorige Woche in Washington war, dem Praesidenten eigenhaendig Ihren Brief uebergeben hat, und dass der Praesident diesen Brief aufmerksam durchlas. Dr. Sachs sprach am vorigen Samstag telephonisch mit seiner Sekretaerin, diese sprach telephonisch mit mir, ich sprach telephonisch mit Wigner und Wigner telephonierte mit Ihnen. Da diese Art der Nachrichtenuebermittlung vielleicht etwas zu indirekt ist,

[27]Our translation.

moechte ich Ihnen heute, nachdem ich Sachs gestern gesprochen habe, brieflich berichten.

Sie haben in Ihrem Brief an Roosevelt den Vorschlag gemacht, dass ein Verbindungsglied geschaffen wird, welches den Kontakt zwischen der Administration und den Physikern aufrecht erhaelt. Roosevelt fragte nun Sachs, welche Form er zur Durchfuehrung Ihres Vorschlages empfehlen wuerde. Dr. Sachs schlug vor, zu diesem Zweck ein Komitee von nicht mehr als drei Personen zu ernennen.

Roosevelt hat diesen Vorschlag akzeptiert und ein Komitee ernannt, bestehend aus Briggs, dem Leiter des Bureau of Standards in Washington, aus einem Colonel, der wohl die Armee repraesentiert, und einem Commander, der, wie ich annehme, die Navy vertritt. Briggs hat an Sachs geschrieben und ihn gebeten, zu der ersten Sitzung dieses Komitees (welche jetzt auf Samstag vormittag festgelegt worden ist), nach Washington zu kommen und Wigner und mich mitzubringen, damit jemand da ist, der ueber technische Einzelheiten Auskunft geben kann. Da es fuer Wigner und mich schwierig sein wuerde, allzu haeufig nach Washington zu fahren, hat Sachs auf meine Bitte Briggs heute telephonisch angerufen und mit ihm verabredet, dass Dr. Teller, der dauernd in Washington ist und den Briggs natuerlich kennt, zu dieser Besprechung zugezogen wird. So werden wir wahrscheinlich nachher ueber Teller den Kontakt mit Briggs auf wirksame Weise aufrecht erhalten koennen.

Falls mich eine etwas hartnaeckige Erkaeltung nicht daran hindert, werde ich, bevor ich nach Washington fahre, noch nach Princeton kommen, und ich hoffe, dass wir dann, wenn es Ihre Zeit gestattet, ueber all die schwebenden Fragen sprechen koennen.

Mit freundlichen Gruessen

Ihr sehr ergebener
Leo Szilard

Document 67

Wardman Park Hotel
Connecticut Avenue & Woodley Road
Washington, D.C.
October 21, 1939

Prof. G. B. Pegram
Physics Department
Columbia University
Broadway at 120th Street
New York City

Dear Professor Pegram:

I wish to give you a short account of yesterday's meeting, at which Briggs acted as chairman. I will give you a longer account in the form of a memorandum, which I am now writing and which I will leave with Briggs before returning to New York. This memorandum is essentially a repetition of the statements and recommendations which I made at the meeting, and it serves the purpose of making things easier for Briggs, when he writes his own report.

On the whole everything came off as could be expected. Teller, who returned from New York, where he spoke with Tuve and Fermi, acted in a double capacity—speaking once in his own name and once in the name of Tuve, who was asked by Briggs to attend the meeting but was unable to come. Tuve put forward certain recommendations which he has discussed and on which he has agreed with Fermi. He said that Government funds ought to be made available for our graphite absorption experiment at Columbia, and named a specific sum, which I do not remember. He also named a sum which he thought ought to be given for purposes of isotope-separation to the University of Virginia, and so on. These recommendations, though they were beside the point, had nevertheless a beneficial effect. The diversion of Government funds for such purposes as ours appears to be hardly possible, and I have therefore myself avoided to make any such recommendation, but Tuve's suggestion provoked detailed discussion of the proposed experiments, and the representative of the Army and the Navy almost committed themselves to the extent of providing some four metric tons of graphite for experiments, if we so desire.

I was astonished how active and enthusiastic Dr. Sachs was during the meeting, and was most favorably impressed. After the meeting he asked me jokingly to confess that I suspected that he was no good, that he would really never get anything done, and that I was surprised, when the time came, that he really became active and started to do things. It seems to me now that he is performing his task efficiently and in the right spirit, and now I am in favor of giving him a fairly free hand, and see what he can achieve.

I expect to be in New York by Tuesday night at the latest.

Yours sincerely,
Leo Szilard

The Briggs Committee failed to produce the sort of organizational and financial support Szilard felt was needed. His hopes for reorganizing the work are explained in the following letters, which also touch on his personal situation.

Document 68

c/o Department of Physics
Columbia University
New York, N.Y.
October 26th, 1939

Dr. Lyman J. Briggs
U.S. Bureau of Standards
Connecticut Avenue
Washington, D.C.

Dear Dr. Briggs:

Enclosed you will find a memorandum[28] in which the statements and recommenda-

---

[28]October 26, 1939; printed in *Scientific Papers*, pp. 204–206.

tions made by me at the meeting of October 21st are repeated and somewhat amplified.

Both at the meeting and in the memorandum I have refrained from putting forward a detailed plan for promoting further research on uranium. Having recently started conversations on this subject with Dr. Pegram, Dr. Fermi, Dr. Wigner and others, I feel that it is best to limit myself to general recommendations until a consensus of opinion on details has been reached.

I personally believe that if sufficient interest in the subject could be aroused, intensive research on uranium might be carried on at four or five different laboratories. Columbia, the Carnegie Institute for Terrestrial Magnetism, the University of Virginia, M.I.T. and Princeton were so far tentatively mentioned in this connection. If a committee, foundation, or some other non-profit organization considered it his task to encourage research on uranium, and had the approval of the government, it could approach the presidents of certain universities in order to obtain the release of some younger physicists from their teaching duties. These men could then devote their entire time to experiments on uranium, which they might want to undertake. They could work either at their own universities, or could work as guests of one of the four or five universities at which larger groups are active on the same subject. In a year or two these men could return to their regular work, and we would thus avoid creating the problem of how to place them later. Such a problem might arise if some of the alternative schemes that have tentatively been put forward were adopted. Also, by proceeding in this way we could avoid interfering with existing research projects in various physics departments, which would inevitably suffer if a large number of men in any single department were persuaded to work on uranium.

One point which might have to be considered in this connection is the following: some of the work which has to be done may be of such nature that the publication of the results had better be avoided. For a young physicist, who has not yet made a name for himself, refraining from publication means a sacrifice which he should not be asked to make without being offered some compensation. Some addition to the salary which he is normally drawing from his university might therefore be desirable and might require the creation of some special fund. This observation is based on experiences gained early in March, when Fermi and I agreed to delay the publication of our experiments on the neutron emission of uranium and attempted to obtain the cooperation of French and English physicists with regard to withholding all publications on this particular subject. I am enclosing for your information copies of the letters and cables exchanged on this issue between February 2nd and April 19th of this year.

Copies of the enclosed memorandum will be sent by me to Dr. Wigner and Dr. Teller, who are old personal friends of mine and with whom I have been in almost constant consultation on this subject since January of this year. I shall also send copies to Dr. Alexander Sachs, Professor G. B. Pegram and Professor E. Fermi. Three additional copies will be sent to you, to be used at your convenience.

Yours sincerely,
Leo Szilard

Document 69

November 5th, 1939

Dr. Alexander Sachs
One South William Street
New York City

Dear Dr. Sachs:

I wish to confirm our appointment for Tuesday night, 7 p.m., at the Men's Faculty Club of Columbia, 400 West 117th Street (117th Street and Morningside Drive). I think you will find both Dr. Pegram and Dr. Fermi very enjoyable persons.

In addition to what I told you over the telephone I should like to make some observations for your personal information:

I expected Briggs to enlarge his committee by including men like you, K. T. Compton or G. B. Pegram. It was a surprise for me to hear that he wanted to include also a group of younger physicists who are themselves actively engaged in doing research on uranium, namely Fermi, Tuve and Beams.[29]

To the inclusion of this second group I should like to make two observations:

1. Since it so happens that the proposed second group includes the name of Fermi we could be assured that the committee will always be well informed and conscientiously advised. The committee would not have to depend on information gathered haphazardously. This may prove to be a very important point and may outweigh all other considerations.

2. The fact that such a second group is being included and that it does not contain my name will make it virtually impossible for me to do in the future what I tried to do in the past, i.e. concern myself beyond the scope of my own experiments with the broader aspects connected with the possibility of a chain reaction, and to act as a driving power in this connection. For me to go on in the future as I did in the past, with a status wholly undefined at a time when some other colleagues have a clearly defined status, would hardly be advisable and in the end probably physically impossible.

I came up against similar difficulties in England six years ago. When the German government started to dismiss German scholars I persuaded Sir William Beveridge to form a committee and create an organization for assisting and placing these scholars. After this was done I went on working for this cause for another six months without having any defined status. Though I finally succeeded in getting a number of things done by exerting myself up to the limit of my strength I learned a lesson, and now I am anxious to avoid a repetition of this experience.

This point may have little importance from a general point of view, but I feel that I have to state my case now so that after the proposed committee has been appointed

---

[29] Jesse W. Beams, University of Virginia, approached for centrifuge separation of uranium-235.

you may not think that I am willfully abandoning a cause when in fact I shall have little choice left in the matter.

In addition to these observations I should like to repeat what I told you over the telephone:

It seemed to me that the omission of the name of G. B. Pegram, who is Head of the Physics Department at Columbia and also Dean of the Graduate School, might be an objective mistake and at the same time also be embarrassing to Fermi. I had a conversation on this subject with Fermi, and we thought that if the committee had the right to co-opt members you might find it perhaps possible to suggest the inclusion of Pegram at the first meeting of the committee.

On Monday I shall telephone your secretary in order to find out if there are any points in the memorandum which you are preparing, or anything else, which you care to discuss with me. I am looking forward to seeing you in any case Tuesday night.

Yours very sincerely,
Leo Szilard

Document 70

December 4, 1939

Dr. Benjamin Liebowitz
350 Fifth Avenue
New York City

Dear Ben:

Early this year I asked you to loan me $2000. (two thousand dollars) in order to make it possible for me and my associates to rent about 1 gram of radium so that we might carry out some very urgent and important experiments on uranium, which we thought might have some bearing on problems of national defense. I told you at that time that I was confident that I could raise some money for these experiments by putting the matter up to some foundation or possibly some department of the Administration, and that I expected to pay back your loan within six months' time. I have spent most of the sum which you loaned me for the renting of 900 mg of radium for a period of ten months. $400. were spent on freeing Dr. Zinn from his teaching duties at City College, and the rest up to the last cent was spent on apparatus which were urgently needed.

Until recently I had hoped that I should be able to obtain financial support for these experiments and be able to return to you the sum which you loaned me. This proved to be impossible, and I had to return the radium and close down experiments. Under ordinary conditions I would naturally refund your loan from my private earnings. Unfortunately, I have not earned anything during this year, as I was tied up with this work on uranium. It looks as though I shall not be able to earn anything next year either, especially since I cannot now return to England. In the circumstances, the prospects of my being able to return to you your loan $2000. are very bad, and I am afraid you will have to consider these $2000. as a bad debt.[30]

I am sorry to have to cause you this disappointment.

Yours very sincerely,
Leo Szilard

---

[30] Years later, Szilard repaid this debt (according to a communication from Liebowitz to G. W. Szilard, July 1, 1964).

Chapter IV
From that point on secrecy was on.

The Washington meeting was followed by the most curious period in my life. We heard nothing from Washington at all. By the first of February [1940] there was still no word from Washington—at least none that reached me. I had assumed that once we had demonstrated that in the fission of uranium neutrons are emitted, there would be no difficulty in getting people interested; but I was wrong. Fermi didn't see any reason to do anything, since we had asked for our money to buy graphite and since we hadn't yet gotten the money. He was interested in working on cosmic rays. I myself waited for developments in Washington, and amused myself by making some more detailed calculations on the chain reaction of the graphite-uranium system.

It is an incredible fact, in retrospect, that between the end of June 1939 and the spring of 1940, not a single experiment was under way in the United States which was aimed at exploring the possibilities of a chain reaction in natural uranium.

Late in January or early in February 1940 I received a reprint of a paper by Joliot in which Joliot investigated the possibilities of a chain reaction in a uranium-water system.[1] In a sense this was a similar experiment to the one which Anderson, Fermi, and I had carried out and published in June. However, Joliot's experiment was made in a different set-up, and I was able to conclude from it what I was not able to conclude from our own experiment, namely, that the water-uranium system comes very close to being chain-reacting, even though it does not quite reach this point.[2] However, it seemed to come so close to being chain-reacting that if we improved the system somewhat by replacing water with graphite, in my opinion we should have gotten over the hump.

I read Joliot's paper very carefully and made a number of small computations on it, and then I went to see Fermi, with whom I was no longer in daily contact because my function at Columbia had ceased. We had lunch together and Fermi told me that he was on the point of going to California. I asked him, "Did you read Joliot's paper?" He said he did. I asked him, "What did you think of it?" and Fermi said, "Not much." At which point I saw no reason to continue the conversation and went home.

I then went to see Einstein again in Princeton and told him that things were not moving at all. I said to Einstein that I thought the best thing I could do now was to go definitely on record that a graphite-uranium system would be chain-reacting, by writing a paper on the subject and submitting it for publication to the *Physical Review*. I suggested that we reopen the matter with the government, and I proposed to take the

---

[1] H. von Halban, Jr., F. Joliot, L. Kowarski, and F. Perrin, *Journal de Physique et le Radium*, ser. 7, *10*: 428–429 (1939).
[2] The French experiment used a much higher ratio of uranium to water than did the American one. The ratio which comes closest to chain-reacting lies between those used in the French and American experiments.

position that I was going to publish my results unless the government asked me not to do so and unless the government was willing to take some action in this matter. (*Docs. 71–72*)

Accordingly, I wrote a paper for publication and sent it to the *Physical Review* on February 16.[3] I brought the paper over to Pegram, who was somewhat embarrassed because Fermi was out of town and Pegram did not know what action he should take. However, he said that he must take some action, so he went to see Admiral Bowen[4] in Washington, who Pegram thought might take some interest because, after all, atomic energy might be used for driving submarines.

On the basis of a conversation which I had with him, Einstein wrote to Sachs, and Sachs wrote again to the President. The President replied that he thought that the best way to continue research would be to have another meeting of the Uranium Committee. (*Docs. 74–79*) And now something most tragic and comic happened. Having received a letter from the White House, Sachs talked to Briggs and suggested a meeting be called. Briggs said he was on the point of calling a meeting and that he wanted to invite Sachs and Dr. Pegram to attend. Sachs said, "Well, what about Szilard and Fermi?" and Briggs said, "Well, you know, these matters are secret and we did not think that they should be included."

Sachs blew up at this point, because this was after all his meeting, and why should the people who were doing the job and who produced the secrets not be included? This, however, was a misunderstanding. Briggs did not want to call the meeting because he had heard from the White House; he wanted to call the meeting at the initiative of Admiral Bowen, whom Pegram had contacted. So Sachs and Briggs talked to each other at cross purposes. They were talking about different meetings. However, somehow things got straightened out and a meeting was called which Fermi and I did in fact attend [April 27, 1940].[5]

In the spring of 1940 we were advised that the money, the $6,000 which the committee had promised us, was available. We bought some graphite and Fermi started an experiment to measure the absorption of this graphite. When he finished his measurement the question of secrecy again came up. I went to his office and said that now that we had this value, perhaps the value ought not to be made public. And this time Fermi really lost his temper; he really thought this was absurd. There was nothing much more I could say, but next time when I dropped in his office he told me that Pegram had come to see him, and Pegram thought that this value should not be published.[6] From that point on secrecy was on. [2]

---

[3]"Divergent Chain Reactions in Systems Composed of Uranium and Carbon." This paper was sent to the *Physical Review* twice, first as a shorter Letter to the Editor on February 6, then in full on February 14 (received February 16), 1940. With each version Szilard sent a covering letter to the editor asking that publication be delayed; it was delayed indefinitely. The paper became Report A-55 of the Uranium Committee, first published in *Scientific Papers*, pp. 216–256.
[4]Admiral Harold G. Bowen, director of the Naval Research Laboratory.
[5]Two following paragraphs have been placed earlier, in chronological order. See n. 8 in chapter III, Recollections.
[6]This value for graphite absorption was very low; had it been published, the German fission project probably would have given more attention to graphite. H. L. Anderson and E. Fermi, Report A-21, in Fermi, *Collected Papers, Volume 2: United States*, 1939–1954 (Chicago: University of Chicago Press, 1965), pp. 31–40.

At this point I received a letter from Turner[7] in Princeton, who pointed out that in the chain reaction which I hoped to be able to set up, there would be formed a new element which might be capable of undergoing fission. As we now know, this is in fact the case, and the element formed in the chain reaction is now called plutonium. Neither Fermi nor I had thought of this possibility, which was obviously of the utmost importance, and this realization increased my sense of urgency. (*Docs. 80–81*)

On Rabi's advice, I enlisted the help of H. C. Urey,[8] who prevailed on the chairman of the Uranium Committee to appoint those of us who were actively interested in this problem to serve as a technical subcommittee of the Uranium Committee. We thought this would put us in a position to approach various laboratories in the U.S. and to enlist their cooperation in pursuing the various aspects of the problem, including the possibility raised by Turner's suggestion. (*Docs. 82–87*)

The Committee, having been duly appointed, met in Washington.[9] When the meeting was opened by the chairman, he told us that the committee would be dissolved upon termination of the current meeting, because if the government were to spend a substantial amount of money—we were discussing sums of the order of a half million dollars—and subsequently it would turn out that it was not possible to set up a chain reaction based on uranium, there might be a congressional investigation. If this were the case, in such a situation it would be awkward if the government had made available funds on the recommendation of a committee whose membership comprised men other than American citizens of long standing. Fermi and I were not American citizens. Though Wigner was an American citizen, he was not one of long standing. Thus the work on uranium in the United States was brought to a standstill for the next six months. Mr. Wigner wrote a very polite letter to the chairman of the Uranium Committee saying that he would hold himself in readiness to work for the government on all matters related to defense, with the exception of uranium. (*Docs. 88–89*)

After reorganization in Washington, which put the Uranium Committee under Dr. Vannevar Bush's committee,[10] Columbia University was given a contract in the amount of $40,000 to develop the Fermi-Szilard system. On November 1, 1940, I was put on the payroll of Columbia University under this contract. Since I was instrumental in inducing the government to assume expenditures for exploring the possibility of setting up a chain reaction, and with a view to the possibility that our efforts might come to nothing, it was deemed advisable to set my salary at a low figure, i.e., $4,000 a year.[11] [*1*] (*Doc. 90*)

---

[7]Louis A. Turner, associate professor of physics at Princeton.
[8]Harold C. Urey, professor of chemistry at Columbia University.
[9]A special advisory group called together by Briggs met at the National Bureau of Standards on June 13, 1940.
[10]Bush's Office of Scientific Research and Development took over uranium work at the end of June 1940. For these and other details of American fission work see Richard G. Hewlett and Oscar E. Anderson, Jr., *The New World, 1939/1946* (University Park, Penn.: Pennsylvania State University Press, 1962), hereinafter referred to as Hewlett and Anderson, *The New World*; and Henry DeWolf Smyth, *Atomic Energy for Military Purposes. The official report on the development of the atomic bomb under the auspices of the United States government, 1940–1945* (Princeton: Princeton University Press, 1945), hereinafter referred to as H. Smyth, *Atomic Energy for Military Purposes*.
[11]This was between the salaries a typical associate professor and a typical full professor would enjoy around 1940.

Documents from February 1940 through October 1940

In early 1940, more than ever convinced that atomic energy was a possibility, Szilard began again to try to encourage self-censorship:

Document 71

King's Crown Hotel
New York City
February 14, 1940

John T. Tate
The Physical Review
University of Minnesota
Minneapolis, Minnesota

Dear Dr. Tate:

Enclosed you will find a manuscript which I am submitting for publication to the Physical Review. There is just a chance that I might be requested by certain departments of the Administration to delay a publication of this paper though I personally do not think that this is very likely to happen. Still if you do not prefer to read the paper yourself, perhaps you might think it advisable to let the referee know about this possibility.

    Pages 22, 23, and 24 of the manuscript deal with two experiments which I described in some detail because it is hoped that it will be possible to give the result of at least one of these experiments within a short time and the values found may be added in proof. Both experiments are in a state of preparation and might be completed by the time the paper appears in print.

Yours very truly,
Leo Szilard

Document 72

King's Crown Hotel
New York City
April 5, 1940

Dear Dr. Tate:

I am writing to you concerning the manuscript of a paper which was sent to you enclosed in my letter of February 14, 1940. I am anxious that this manuscript should not be sent to print until I have definitely heard from the Administration that there is no objection to its publication. In the meantime, however, I should be glad to know whether the manuscript has been accepted for publication in the Physical Review and perhaps you would be kind enough to inform me with regard to this point.

Yours very truly,
Leo Szilard

Document 73

King's Crown Hotel
New York City
April 12, 1940

Professor F. Joliot
Collège de France
Paris, France

Dear Professor Joliot:

Many things have considerably changed since March last year, and therefore I should like to raise once more the question whether or not results concerning chain reactions in uranium ought to be published.

It is reported that such publications are prevented in Germany and that work on uranium there is carried out in secrecy.

As you know, my own inclination would be to delay publications on this subject, but I have not discussed this matter with anyone else here in America since April last year, and I do not know what view others would take if the question were to be raised again. If, however, I should hear from you that in the meantime you have adopted some new policy of delaying publications, I could then perhaps talk to others here and find out what the general feeling is on this subject.

Everybody here hopes that you and your colleagues will not suffer too many inconveniences on account of the war and that your work will go on uninterrupted.[1]

With best wishes to all,

Yours sincerely,
Leo Szilard

Essentially no progress had been made on fission over the winter, so Szilard mobilized Einstein and Sachs to make another approach to Roosevelt:

Document 74

King's Crown Hotel
New York City
March 7, 1940

Dear Professor [Einstein]:

I have sent Dr. Sachs the draft of the letter we discussed together. As you see from his

---

[1] We have found no reply from Joliot and we are uncertain whether he received this letter. His group kept secret all the work they did after the war began in Europe (September, 1939).

enclosed letter, he suggests certain changes. These and other changes are entered in pencil in the last draft and you can see from the scribble which suggested changes are Sachs' and which are mine.

In the enclosed folder you will find the new version in which I have tried to fulfill the wishes of Dr. Sachs as far as possible. There is a copy for your files in the folder.

Should you wish to change back any of the alterations or anything else, you could mark your corrections in the clean copy and return it to me for retyping.[2]

With friendly greetings,

Yours very sincerely,
Leo Szilard

Document 75

A. Einstein
112 Mercer Road
Princeton, N.J.
March 7, 1940

Dr. A. Sachs
c/o Lehman Corp.
1 South William St.
New York, N.Y.

Dear Dr. Sachs:

In view of our common concern in the bearings of certain experimental work in problems connected with national defense, I wish to draw your attention to the development which has taken place since the conference that was arranged through your good offices in October last year between scientists engaged in this work and governmental representatives.

Last year when I realized that results of national importance might arise out of the research on uranium, I thought it my duty to inform the Administration of this possibility. You will perhaps remember that in the letter which I addressed to the President I also mentioned the fact that C. F. von Weizsaecker, son of the German Secretary of State, von Weizsaecker, was collaborating with a group of chemists working upon uranium at one of the Kaiser Wilhelm Institutes, namely, the Institute of Chemistry. Since the outbreak of the war, interest in uranium has intensified in Germany. I have now learned that research there is being carried out in great secrecy and that it has been extended to another of the Kaiser Wilhelm institutes, the Institute of Physics. The latter has been taken over by the Government and a group of physicists, under the leadership of C. F. von Weizsaecker, who is now working there on uranium in

---

[2]Our translation from the German.

collaboration with the Institute of Chemistry.³ The former director was sent away on a leave of absence apparently for the duration of the war.

Should you think it advisable to relay this information to the President, please consider yourself free to do so. Will you be kind enough to let me know if you are taking any action in this direction.

Dr. Szilard has shown me the manuscript which he is sending to the Physics Review in which he describes in detail a method for setting up a chain reaction in uranium. The papers will appear in print unless they are held up, and the question arises whether something ought to be done to withhold publication. The answer to this question will depend on the general policy which is being adopted by the Administration with respect to uranium.

I have discussed with Professor Wigner of Princeton University and Dr. Szilard the situation in the light of the information that is available. Dr. Szilard will let you have a memorandum informing you of the progress made since October last year so that you will be able to take such action as you think in the circumstances advisable. You will see that the line he has pursued is different and apparently more promising than the line pursued by Monsieur Joliot in France about whose work you may have seen reports in the papers.⁴

Yours sincerely,
Albert Einstein[5]

Document 76

[Sachs]
March 15, 1940

Dear Mr. President:

As a sequel to the communication which I had the honor to submit to you on October 12, Professor Albert Einstein sent me another regarding the latest developments touching on the significance of research on uranium for problems of national defense. In that letter he suggests that I convey to you the information that has reached him that since the outbreak of the war, research with uranium is being carried out in great secrecy at the Berlin Institute of Physics, which has been taken over by the Govern-

---

[3]This was true, and the Berlin group was also collaborating with several other groups concerned with fission. See David Irving, *The German Atomic Bomb. The History of Nuclear Research in Nazi Germany* (New York: Simon & Schuster, 1967).
[4]Presumably a reference to Szilard's choice of graphite as a moderator; in their last published papers the French were still studying uranium-water systems, and announced that they had produced a non-self-sustaining chain reaction. See Halban, Joliot, Kowarski, and F. Perrin, *Journal de Physique*, ser. 7, *10*: 428–429 (1939).
[5]A copy of this letter is in the Szilard files. It was also published in *Einstein on Peace* (New York: Schocken Books, 1968), pp. 299–300. Reprinted by permission of the Estate of Albert Einstein.

ment and placed under the leadership of C. F. von Weizsaecker, son of the German Secretary of State.

In the realization that these further views of Dr. Einstein have a definite bearing on the favorable report submitted to you by Dr. Briggs as Chairman of the Committee which conferred with experimental scientists concerned and myself, I am enclosing his communication for your kind perusal. May I also ask whether and when it would be convenient for you to confer on certain practical issues brought to a focus by the very progress of the experimental work, as indicated in the concluding paragraph of Dr. Einstein's letter.

In view of your original designation of General Watson[6] in this matter, I am transmitting it through his good offices.

Yours sincerely,
[A. Sachs]

The President
The White House
Washington, D.C.

Document 77

April 5, 1940

The White House
Washington

My dear Dr. Sachs:

I am grateful for your letter of March fifteenth enclosing the information from Dr. Einstein regarding the recent development in Uranium research. I have asked my Secretary, General Watson, to arrange another meeting in Washington at a time convenient for you and Dr. Einstein. I think Dr. Briggs should be included, and special representatives from the Army and Navy.

I am of the opinion that this is the most practical method of continuing this research, and I shall always be interested to hear the results.

Very sincerely yours,
Franklin D. Roosevelt

Dr. Alexander Sachs
One South William Street
New York, N.Y.

On April 15, 1940, Sachs wrote to Einstein that organizational work on nuclear fission had resumed.[7] Ever since Szilard had first acquainted him with the problem, said Sachs, he had been in continuous touch with him at every stage of developments. Sachs expressed

---

[6]Edwin M. Watson, secretary to President Roosevelt.
[7]Copy in Szilard files.

the hope that Einstein, like Wigner and Szilard, would plan to attend the new meeting of the Briggs Committee. The following documents show the ideas Szilard was expressing to Sachs and Einstein's response.

Document 78

King's Crown Hotel
New York City
April 22, 1940

Dr. Alexander Sachs
c/o Lehman Corporation
One South William Street
New York City

Dear Dr. Sachs:

In accordance with the letter written to you by Professor Einstein on March 7, I am submitting to you in the following a memorandum dealing with the present work on nuclear chain reactions. Only one aspect of the subject is discussed in this memorandum, namely its possible bearing on questions of national defense.

Memorandum

We have to discuss separately two different types of chain reactions, i.e.

a) chain reactions in which the neutrons are slowed down, and in which only a small fraction of the uranium can be utilized, corresponding to the content of uranium-235 in ordinary uranium; (if ordinary uranium is used for the purposes of such a chain reaction, a ton of uranium will be exhausted after having supplied as much energy as corresponds to the burning of about 3000 tons of oil);

b) chain reactions in which the neutrons are not slowed down and in which the bulk of the ordinary uranium could be utilized; (if it were possible to maintain a chain reaction of this type in uranium, one ton of uranium could supply more energy than 300,000 tons of oil).

There is reason to expect that a chain reaction of the type described under a) can be maintained in a system composed of uranium and carbon.

Whether or not a chain reaction of the second type, as discussed under b), can be maintained in uranium is not known and has for the present to be considered an open question which, in view of its far reaching consequences, urgently requires further study.

PART I
CHAIN REACTIONS MAINTAINED IN SYSTEMS COMPOSED OF
CARBON AND URANIUM

A chain reaction of this type is capable of applications which may have a bearing on questions of national defense.

1. A system composed of carbon and uranium might be used for purposes of power production. Questions relating to the transformation into power of the energy liberated in the chain reaction as well as questions relating to the regulation of the chain reaction have been studied, and methods for solving these problems have been devised.

Personnel has to be protected from being exposed to the radiations emanating from the chain reaction by means of water tanks, and such an atomic engine equipped in this way could be used as a power reserve in larger naval units. The weight of the water tanks rules out the possibility of using an atomic engine for the purpose of driving aeroplanes.

One ton of uranium would be capable of supplying about as much power as 3000 tons of oil. For instance, a 30,000 ton battleship, which would ordinarily have a maximum oil load of 4000 tons could in the future be equipped for the use of both oil fuel and atomic power and would carry perhaps 1000 tons of oil and 50 tons of uranium, the latter representing the equivalent of an oil reserve of about 150,000 tons. Accordingly, such a boat would have a practically unlimited cruising radius.

Since a battleship equipped with an atomic engine need not carry in war-time more than a normal oil load of perhaps 1000 tons, there would result a saving in weight, even if allowance is made for the weight of the atomic engine. This saving in weight would lead to an increase in the top speed of the vessel.

The limited supply of uranium would make it inadvisable to use up any considerable amounts for naval purposes in peace time, and the atomic engines with which battleships may be equipped must not be used except occasionally in maneuvers and in case of actual warfare. Since a large battleship or battlecruiser will use more than 1/2 ton of oil per mile if cruising at an economical speed, it would exhaust its full oil load of about 4000 tons during a cruise covering about 10,000 miles. This means that a fast ship can not operate for any length of time at a distance of about 4–5000 miles from its nearest base.[8] The advantage of a battleship having an equivalent of an oil reserve of 150,000 tons would in these circumstances be decisive, since apart from the increased speed it could stay for a long period near its objective at any distance from its base.

2. A system composed of carbon and uranium may be used as a weapon in the following manner: A chain reaction may be maintained in this system and the neutrons emanating from the chain reaction may be allowed to escape. The intensity of the neutron radiation could be made so high that this radiation would fatally injure by its physiological action human beings who are exposed to it within a radius of one kilometer. By mentioning this fact it is not desired to imply that such a system represents a desirable or particularly efficient military weapon.[9] The reason for emphasizing this point lies rather in the belief that such a system could be used as a weapon by some other country during the present war, possibly in the near future,

---

[8]Szilard (who relied on others for his information on naval strategy) has not taken refueling at sea into account.
[9]Szilard's proposed device, a very large unshielded nuclear reactor, would indeed be a clumsy weapon.

and that it could be used with considerable effect on a country which is not prepared to meet this new type of attack.

PART II

CHAIN REACTIONS IN WHICH THE NEUTRONS
ARE NOT SLOWED DOWN[10]

It is not known at present whether or not chain reactions of this type can be brought into existence. If, however, this could be done they would have a bearing on questions of national defense, going in their scope of applications far beyond the applications discussed in Part I.

1. In a chain reaction of this second type one ton of uranium used as driving power in a warship could supply more power than 300,000 tons of oil. Consequently, it would probably be possible for the larger types of naval vessels to dispense entirely with the use of oil.

2. A chain reaction of this second type would make it possible to bring about explosions of extraordinary intensity. If for purposes of aggression, a bomb based on such a chain reaction were set off at sea near the coast, the tidal waves brought about by the explosion might lead to the destruction of coastal cities. Such a bomb would not be too heavy to be carried by small boats, but could hardly be carried by existing airplanes.

Yours sincerely,
Leo Szilard

Document 79

[Einstein]
April 25, 1940

Dr. Lyman J. Briggs, Director
National Bureau of Standards
U.S. Department of Commerce
Washington, D.C.

Dear Dr. Briggs:

I thank you for your recent communication concerning a meeting of the special Advisory Committee appointed by President Roosevelt.

As, to my regret, I shall not be able to attend this meeting, I have discussed with Dr. Wigner and Dr. Sachs particularly the questions arising out of the work of Dr. Fermi and Dr. Szilard. I am convinced as to the wisdom and the urgency of creating the conditions under which that and related work can be carried out with greater speed

---

[10]Here Szilard considers a fast-neutron chain reaction in natural uranium, which in fact is impossible. He is not yet aware of the feasibility of obtaining significant amounts of uranium-235 or plutonium, substances in which the fast-neutron chain reaction does work.

and on a larger scale than hitherto. I was interested in a suggestion made by Dr. Sachs that the Special Advisory Committee submit names of persons to serve as a board of trustees for a non-profit organization which, with the approval of the Government committee, should secure from governmental or private sources, or both, the necessary funds for carrying out the work. It seems to me that such an organization would provide a framework which could give Drs. Fermi and Szilard and co-workers the necessary scope. The preparation of the large scale experiment and the exploration of the various possibilities with regard to practical applications is a task of considerable complexity. I think that given such a framework and the necessary funds, it could be carried out much faster than through a loose cooperation of University laboratories and Government departments.[11]

Yours sincerely,
Albert Einstein

Szilard's continual attempts to organize secrecy finally found a response, and by early June the community of physicists imposed censorship on fission work. The following letters trace Szilard's efforts.

Document 80

[Turner]
Palmer Physical Laboratory
Princeton University
Princeton, New Jersey
May 27, 1940

Dr. Leo Szilard
Pupin Physics Laboratories
Columbia University
New York, New York

Dear Szilard:

Enclosed is a copy of the manuscript of a Letter to the Editor on the subject of fission.[12] I thought that you would be interested in it. Wigner tells me that some of the work on the subject is not being published at present because of its possible military value. I find it a little difficult to figure out the guiding principle in view of the recent ample publicity given to the separation of isotopes. Nevertheless, if that is the case, I should be pleased if you would turn this over to the authority on such matters and I shall be glad to hear from him and conform to his wishes. It seems as if it was wild

---

[11] A copy of this letter is in the Szilard files. It was also published in *Einstein on Peace*, pp. 300–301. Reprinted by permission of the Estate of Albert Einstein.
[12] This manuscript raised the theoretical possibility of fission in element 94 (later named plutonium), which can be created in a uranium nuclear reactor. Turner probably did not yet realize that plutonium could be used to make a bomb.

enough speculation so that it could do no possible harm, but that is for someone else to say.

Wigner also spoke about some general plans which are developing for a large scale concerted attack on this problem of getting atomic energy out of uranium. He thought that perhaps we could do something about it here. I should be very glad to assist in that enterprise if there is any useful part that I could play in it. I do have a few ideas as to methods of attacking the problem. Naturally I don't just want to charge ahead and start some research which some of the rest of you have probably considered and rejected or considered and planned to begin. We'll just let the matter rest until we hear something from you further. I was sorry that I didn't have a chance to talk to you at some length the other day when you were down.

Sincerely yours,
Louis A. Turner

LAT:MH
Enclosure

Document 81

King's Crown Hotel
New York City
May 30, 1940

Professor Louis A. Turner
Palmer Physical Laboratory
Princeton University
Princeton, New Jersey

Dear Turner:

I am very grateful to you for letting me have a copy of your manuscript which might eventually turn out to be a very important contribution.

You are certainly justified in finding it difficult to figure out the guiding principle which regulates at present what is being kept secret and what is not. However, things are perhaps not as bad in this respect as they might seem and, at any rate, a sincere effort is being made to bring order out of chaos. The publicity given to the separation of isotopes is rather unpleasant and was regretted by all those with whom I collaborate, but at present there is a view that we may now make the best of it by using it as a smoke screen behind which other work might go on in comparative seclusion.

As you perhaps know, I have written a rather detailed paper on the subject of chain reactions which was sent to the Physical Review early in February but I have been asked to delay the publication of this paper and to refrain from discussing the subject matter for the time being.[13] This was the reason why I did not feel free to show you more than those few pages in which you had "legitimate" interest.

---

[13] In fact it was Szilard himself who made this request.

Obviously, we are at present in an awkward situation which requires a better adjustment. It appears important that free discussion of all results and ideas among as many physicists as is practicable should not be inhibited and I believe that it is our right and duty to insist that such free discussion should not be hindered by undue secrecy. Perhaps the best solution would be to draw up a list of all trustworthy people who wish to do serious work on uranium and to have free discussion within this group. An uncontrolled diffusion of information would be prevented by pledging those included in this list to refrain from discussing the subject with those who are not included in the register. From time to time new names could be added as the need arises. Manuscripts, the publication of which is being delayed, would be communicated to everybody within the group. I have the impression that some solution of this type will be worked out in the near future and you will be approached as soon as such a solution is worked out.

At the last meeting at which this subject was discussed a representative of the Government suggested that the scientists might themselves form some sort of voluntary association and impose upon themselves the restrictions concerning publications which appear to be necessary in order to safeguard the required secrecy. Professor Urey has now taken upon himself the task of carrying out this suggestion and he will have a discussion on this subject with the Government authorities in the next few days.

In the circumstances I felt that the best course for me to take was to hand over your letter to Urey rather than send your manuscript to the Government departments concerned. By choosing this avenue it will take longer for you to hear officially anything about the fate of your paper, but on the other hand, we take less risk in the long run that our work will be hampered by undue secrecy.

In the meantime, you could perhaps write to Tate advising him that your paper is being submitted to certain Government departments and ask him to delay the publication until he hears from you to the contrary.

From what I know there is little doubt that the publication of your paper will have to be delayed indefinitely in the same way as that of my own last paper.

If you wish me to do so I could transmit your paper direct to the Government departments interested and ask point-blank for a decision in this particular case. However, if it is agreeable to you, I would rather await the outcome of Urey's discussion with the authorities and then have your paper submitted by Urey.

Your paper is certainly very stimulating even if somewhat hypothetical and I was very glad to have an opportunity to read it. As I repeatedly explained to Wigner I personally would be very happy if you at Princeton could collaborate with the rest of us and I shall get in touch with you as soon as I am free to do so. If there is no other solution I might get in touch with you in Woodshole[14] and perhaps run up for a day if there is anything important to settle before you return. We could then discuss things in greater detail. Could you possibly let me have your Woodshole address?

---

[14] i.e., Woods Hole, Mass.

Please consider all the information contained in this letter as confidential, and I should be very grateful if you did not discuss it with anyone except Wigner to whom I am sending a copy.

Could you possibly confirm whether you have asked Tate for a temporary delay until further notice by dropping me a line?

Yours sincerely,
Leo Szilard

Szilard continued his efforts to organize a relatively painless self-censorship and also to create a group of effective leaders for the uranium project:

Document 82

Memorandum for Professor Urey May 30, 1940

Admiral Bowen suggested at a meeting held under the chairmanship of Dr. Briggs at the Bureau of Standards on April 27, 1940, that the scientists working on uranium should form some sort of voluntary association and impose upon themselves such limitations concerning the publication of results as appears to be necessary.

While at the time this suggestion was made it seemed to be difficult to get the co-operation of the majority of scientists, the invasion of Holland and Belgium has brought about a change of attitude so that now we may hope to succeed if we act on the suggestion of Admiral Bowen.

It is proposed that a committee for the "coordination of nuclear research" be formed under your chairmanship and that this committee formulate from time to time the policy which is to be adopted with regard to publication. If this committee were composed of yourself, Pegram, Wigner, Beams, Tuve, Teller, Fermi, and myself, it would be easy to meet once a month and to deal with all problems which may arise. For this reason no names have been included from the Middle-West or the West coast. Since, however, the Physical Sciences Division of the National Research Council has appointed a committee for the purpose of looking into the question of uranium which consists of Beams, Breit,[15] and Pegram, you might feel that you want to ask Breit to join the committee so that all members of the group representing the National Research Council should be included in your committee.

Your committee could have a sub-committee for unseparated uranium and a sub-committee for the separation of uranium isotopes. Fermi and I would be glad to act as secretaries to the sub-committee for unseparated uranium and I suppose you and Beams might be willing to act as secretaries for the subcommittee for the separation of uranium isotopes.

[15]Gregory Breit, professor of physics, University of Wisconsin. This committee was formed at his urging, as a result, he tells us, of conversations he had with Wigner and particularly with Szilard.

The scope of the committee could be enlarged immediately after its formation by including the non-governmental members of the Special Advisory Committee which has been meeting under the chairmanship of Dr. Briggs. These non-governmental members are Professor Pegram, Dr. Alexander Sachs, and Professor Albert Einstein. They, together with yourself, could then form the link between your committee and the government and could act as a nucleus for a board of trustees. Such a board of trustees will be required if funds are to be obtained or solicited from either governmental or private sources.

In order to be able to maintain the necessary secrecy and at the same time to preserve the possibility of free discussion among those scientists who wish to cooperate with each other, it is proposed that your committee after its formation should draw up a list of names and that there should be free discussion among those who are included in this register. At the same time, an uncontrolled diffusion of information would be prevented by pledging all those to be included in this register to refrain from discussing the subject of uranium with anyone else. New names could be added to the list from time to time in order to include all those who are trustworthy and who may wish actively to collaborate. Separate lists of names may be drawn up for the various branches of uranium research in accordance with the fact that the need for secrecy is greater for some branches than for others.

Requirement For Funds

Fermi and I would desire to carry out a large scale experiment which would involve the use of about 100 tons of graphite and 10 to 20 tons of metallic uranium. Before actually placing orders for such an experiment which will involve considerable expenditure we propose to go through a preparatory stage involving an expenditure of $50,000. The successful completion of this preparatory stage would make it possible to carry out the large scale experiment in a comparatively short time and with an increased assurance of success.

We are looking forward to obtaining from the Government the sum of $50,000. which is required for this preparatory stage. We feel, however, that a few weeks or months may pass before we will be actually in the position of making financial commitments on the basis of the expected action by the Government. Unless we are able to make such commitments within the next two weeks up to the amount of $15,000. we shall not be able to efficiently prepare the work which otherwise could be speedily carried out during the summer and during the next academic term. This means that we may lose four to six months of valuable time. If this amount could be obtained without delay from a private source, for instance, from the Carnegie Institute through Dr. Bush, it would represent a very great help at this juncture. It could be either refunded if and when Government facilities become available or it could be handed over to your committee earmarked for work on unseparated uranium and used for such expenditure as will not be provided for by the Government.

Of the required $15,000. about $12,000. might be taken up for assuring the collaborators whose help we need adequate salaries for a period of a year. We propose to keep the salaries somewhat higher than usual in order to compensate our collaborators for the damage which their careers will suffer by their being prevented from publishing any papers.

While undoubtedly a fund of $25,000. would be preferable inasmuch as it would include an item of $7000. for buying materials such as uranium oxide and uranium metal in quantities required within the next six months and another item of $6,000. for building apparatus, we feel that if we could be sure right away that we can go ahead and make commitments on the basis of a budget of $15,000. this smaller sum would be sufficient to bridge the gap provided that we receive a pledge by the Government concerning the budget of $50,000. by the end of September.

Document 83

Memorandum for Dr. Sachs

Please find enclosed memorandum for Dr. Urey of May 30, 1940. In addition to the items included in the above-mentioned memorandum the following points seem to require attention.

a. It is important that Dr. Urey and the non-governmental members of the Special Advisory Committee be authorized to investigate whether there is a possibility of mining uranium ore in the Belgian Congo and transporting it to this country under the present conditions. If it is considered premature for the Government to buy any uranium ore perhaps some arrangement could be made with Dr. Sengier, the managing director of the Union Miniere who is at present in New York, or through the Belgian Government in exile that uranium ore be brought to the United States with the assistance of the United States Government, the Belgian company retaining the title of this ore but committing itself not to re-export it without special permission. It is impossible to know whether such and other alternative solutions are feasible, unless a preliminary inquiry is made, and it is not advisable to make such an inquiry without proper authorization.

b. It appears necessary that some experimentation be started at once by industrial firms who are willing to supply 10 to 20 tons of uranium metal at about six months notice. It is necessary that the non-governmental members of the Special Advisory Committee and Dr. Urey should be in a position of approaching industrial firms on this subject and should feel authorized to do so.

c. It would be desirable that Dr. Urey and the non-governmental members of the Special Advisory Committee should form the nucleus for a board of trustees and work out the statutes for some non-profit organization which would act as a link between the Government and the university laboratories. If such an organization were formed the physicists ought to be encouraged to take out patents for their inventions which would be assigned either to this non-profit organization or direct to the Government. In any case the Government would thus be safeguarded against having to pay royalties for the use of such inventions, which otherwise might be patented by industrial firms whose research employees begin to show increasing interest in this field of development.

In this connection the question has to be raised whether it is possible to keep such patents secret. In order to do so in an adequate way it might be advisable to modify the present law if necessary. Such a modification of the present law ought of course not to be made exclusively with a view to inventions concerning chain reactions but also with a view to all inventions which have important applications in national defense. The physicists and engineers ought not to be deprived of the stimulus arising out of the possibility of patenting their inventions and at the same time collaborating with the Government in the effort to keep certain of these inventions secret.

Document 84

[Turner]
Palmer Physical Laboratory
Princeton University
Princeton, New Jersey
June 1, 1940

Dear Szilard:

Your letter of May thirtieth was received. It seems to me that the present situation is a very unsatisfactory one. If the matter is really important, it should not be on such a catch-as-catch-can basis. Assume for the sake of argument that my paper is a really important contribution. The question of its being published or held up should not have to depend on the accident of our being acquainted and my having sent it to you. There ought to be some general way in which the thing was being handled—right now, not week after next or some later time.

I find it hard to understand how it can be that one important paper was held up in February and now, more than three months later, the matter is in the stage of being discussed informally sometime next week. I think that it is better that I decline to ask for delay in the publication of my paper. Please do not misunderstand me. I am not anxious to rush publication for any personal reasons, and I certainly do not want to fail to cooperate reasonably. I feel that there is a matter of principle involved; that it is high time that the matter was brought to a focus; that somebody, either in the government or outside, like Urey, should take the authority and request all editors to defer publication of papers on the subject until some plan has been worked out. I feel that if this minor paper can produce some action instead of talk it may be of some importance quite apart from any ultimate consequence of the ideas expressed in it.

I am sending a copy of this letter to Tate and also one to Urey.

Sincerely yours,
Louis A. Turner

P. S. Wigner and I will see you Monday. The above is my present reaction. I feel that something should be done right away, details can be worked out later.

Document 85

[Breit]
The University of Wisconsin
Madison
Department of Physics
June 5, 1940

Dear Szilard:

I have received from Tate Turner's Letter to the Editor and a copy of Turner's letter to you. I do not know what you have written to Turner but it appears that you do not think the letter should be published or at least that it should be delayed. I should like to know your opinion and Fermi's very much indeed.

As chairman of the committee of the Division of Physical Sciences on uranium fission I have written Tate concerning the advisability of control over such publications. Tate is very willing to cooperate and the present understanding is that he will send me all such papers so that the committee may decide on whether they should be published. The committee consists of Pegram, Beams and myself. I have also asked for Wigner to be appointed. It may be best for you and Fermi not to be officially on the committee but I plan, of course, to have the benefit of your advice.

I should suggest that Fermi speak to Urey asking for control of publications in the Journal of Chemical Physics and in the publications of the American Chemical Society. I have asked for such control through official channels but there are unavoidable delays.

Sincerely yours,
G. Breit.

Szilard remained very concerned with the organization of the fission work, as is indicated in the following letters, culminating in a memorandum which gives a sharp critique of the state of affairs and then sketches a vision of possible future scientific organization:

Document 86

King's Crown Hotel
New York City
July 4, 1940

Dear Fermi:

I think I ought to write you about the following events which took place after you left:
 1. Lawson,[16] after careful consideration, decided that he would prefer to go to the

[16]Andrew W. Lawson, Jr., who received his Ph.D. from Columbia in 1940.

University of Pennsylvania next fall as previously arranged. Since I told him that I would try to get a salary of $3000. per year for him, his decision was not due to a lack of financial inducement.

2. Wigner told me during the last meeting which we jointly attended in Washington that he intends to withdraw from further cooperation with the Government representatives on the subject of uranium. At my request he refrained from saying anything about this during the meeting but I understand that he wrote a letter to this effect a few days later to Urey and Briggs. I do not think that we ought to worry unduly about this since I am sure that if a proper frame-work is eventually created for carrying out work on uranium it will be possible to get Wigner's cooperation. For the present though his resignation is one of several disturbing elements.

3. Bowen has shifted his grounds insofar as he now prefers to give the university a lump sum rather than pay salaries and everything else as much as possible directly to the recipient. Moreover, according to the present plans, Bowen will support isotopic separation rather than the work we contemplated and our project is supposed to receive support through another Government committee which is headed by Bush. I assume that Professor Pegram will write you about this in greater detail so I need not go into this for the present.

4. I compiled an estimate of cost for a complete survey of the nuclear constants involved which you will find enclosed. Though the experiments eventually carried out may be very different from those which I have listed the changes will hardly affect the conclusion that about $50,000. will have to be spent by the time the survey is carried out. I believe that you share my opinion that such a survey has to be carried out whether or not the intermediate experiment shows a positive result. According to present plans, $90,000. would be requested for buying materials for the intermediate experiment and I believe our policy should be to give the intermediate experiment the right of way before the general survey of nuclear constants. I feel that this will be a somewhat academic statement since we may have to wait for quite a long time before we get materials for the intermediate experiment.

5. Enclosed you will find a copy of a letter which I sent to Gunn which speaks for itself.

6. Sachs and Urey saw the Belgians. From what Sachs tells me we have probably "missed the bus" as far as Belgian ore is concerned. As far as uranium oxide is concerned he has the impression that the Belgians will treat it merely as a business matter and if they are not handled skillfully they will charge an exaggerated price for the amounts which we need for the large scale experiment. In these circumstances, it appears essential to find out from the Canadians just how much they are able to do for us and Professor Pegram is looking after this end of the matter.

7. During the last fortnight I was considerably worried by doubts concerning the possibility of realizing my conception of our collaboration within the frame-work of the Physics Department. Having discussed with you my conception of our proposed collaboration quite extensively in March and also shortly before you left for Chicago

I considered this point settled and it was not my intention to raise in this connection any questions of principle. However, it so happened that the question came up more or less accidentally in a conversation with Professor Pegram and was raised by him rather than by me.

Having explained to Professor Pegram what I had previously explained to you I believe he has now a clear picture of the stand which I propose to take with regard to the question of principle which is involved. At first I thought that we might leave the matter open until your return though this did not appear necessary since there are no questions of detail which will require being discussed. On further consideration I felt that it may be better to ask Professor Pegram to take any decision which may be required in this matter as soon as possible, and no doubt he will want to consult you before doing so. You may therefore expect to hear from him in the course of the next few days in this connection.[17]

With kind regards.

Yours,
Leo Szilard

Document 87

King's Crown Hotel
New York City
July 6, 1940

Dear Breit,

Many thanks for your letter.[18] Following the conversation we had on our way from Washington to New York, I have given some thought to the issue mentioned in your letter and I am now entirely convinced to your point of view. Consequently, I am taking a strong stand in favor of an experiment on as large a scale as possible. This large scale experiment, or some intermediate experiment, operating with at least five tons of uranium ought to have the right-of-way before the general survey of the nuclear values involved. Nevertheless, this general survey will also have to be carried out.

There is another point about which I became converted to your opinion. I now think that steps should be taken to prevent certain publications in Nature and the Proceedings of the Royal Society of London. With the collapse of France there is an immediate danger that Joliot and his co-workers will start publishing something of their previous work in these periodicals.

---

[17]On June 19, Szilard drafted but did not send a letter to Fermi, proposing that they collaborate on equal terms with each empowered to carry out experiments independently. In the end Fermi took charge of experimental work while Szilard provided ideas and worked on the procurement of pure materials.

[18]Breit to Szilard, June 20, 1940, saying, "I still think that more rapid progress will be achieved by arranging an intermediate or full scale experiment rather than by careful measurement." Fermi believed otherwise.

On the other hand I feel even more strongly than before that your attempt to prevent publication will break down unless we create a satisfactory substitute in the form of some private publication. If that is not done there will be a growing tendency towards indulgence and finally practically everything will be published as it has been in the past. I wonder whether you have given the matter further thought since your return to Madison.

With kindest regards.

Yours,
Leo Szilard

Document 88

King's Crown Hotel
New York City
July 6, 1940

Professor E. Wigner
c/o Physics Department
University of Michigan
Ann Arbor, Michigan

Dear Wigner:

I am enclosing a letter which I just had from Polanyi and which does not sound very cheerful.[19] I am embarrassed about answering his question since I believe that no information should be sent abroad. However, I shall perhaps try to write up something and give it to Urey with the request of passing it on if Urey is willing to take the responsibility.

I now believe that your resignation was exactly the right thing to do and that it will have some beneficial effect. Urey did not understand your letter but Briggs did. I hope that eventually some framework will be set up for the work on uranium and that I shall then be in the position of persuading you to collaborate, but for the time being there is an increasing amount of confusion and a constant change of the personnel of committees concerning uranium and a growing dissatisfaction on my part as well as on the part of Sachs. I almost reached the point of following your example. I had a growing suspicion during the last fortnight that Pegram's conception concerning the role which I am going to play in the work is very far from my own conception. Finally I decided to explain to him the stand which I propose to take in connection with the principle which is involved and ask him to take any decision which may be required in this connection as soon as possible. Naturally he will consult Fermi before doing so. I am writing you on this in order to keep you informed but I do not think that you ought to intervene in any way. Sachs is very much aroused by the way in which matters are treated in Washington on the part of Briggs and for the present I have to restrain

[19]Polanyi (University of Manchester, England) to Szilard, June 18, 1940, asking for information about work on uranium-235 separation in the United States.

him from taking the matter up with the White House. As far as Wheeler[20] is concerned I do not think that Breit or you need to worry about this aspect for the present. I am keeping him constantly in mind and I believe I can convince Pegram very soon that only part of the necessary experiments can be done at Columbia and that a collaboration with other universities is essential. I am enclosing a copy of a letter I wrote to Wheeler some time ago to show you the line which I am taking in the meantime.

I shall be very glad to see Dr. Torda if I hear from her.

Yours,
Leo Szilard

Document 89

August 28, 1940

Dr. Alexander Sachs

Dear Dr. Sachs:

Enclosed you will find a rough draft which I am certain you will want to change in many places.

I am sending it to you in advance of my visit which is scheduled for 5 p.m., so that you may be able to read it if you wish to do so before we discuss it orally.

Yours very sincerely,
Leo Szilard

[Enclosure in preceding letter]

August 27, 1940

In September last year I was approached by a group of scientists with the request of helping them to enlist the support of the Government for a line of work which might be of great importance for the U.S. Navy. Thereupon the attention of the Administration was drawn to this line of research and its potential possibilities by a letter written by Professor Albert Einstein, which was addressed to the President and which I transmitted to him in a personal interview. . . . [21]

3. Shortcomings of the present status concerning the organization of the line of work pursued by Dr. Fermi and Dr. Szilard

The production of a chain reaction in uranium in circumstances which are suitable to be utilized in an engine capable of driving a naval vessel is a task of considerable com-

---

[20]John Wheeler, physicist at Princeton. Szilard's letter to him (June 20, 1940) discussed calculations and measurements of the behavior of neutrons in a reactor.
[21]We omit five paragraphs describing the current organization of the work and the value of uranium for propelling warships.

plexity. This task cannot be carried out by a loose cooperation of various committees and universities. It requires planning, the preparation of experiments six months and occasionally one year ahead, the gathering of a group of physicists prepared to collaborate for a number of years, whose loyalty has to be with this work rather than with the individual teaching institutions with which they happen to be associated. At present we may assume that the work carried out at Columbia University will be supported by Dr. Bush's committee, but it is unlikely that this complex task can be carried out at a single university where the available space and the necessity of maintaining the routine work carried out by the department will naturally limit the speed at which the work should be carried out. A loose cooperation between various universities can be anticipated and may to some extent remedy the situation, but surely this is no way of obtaining quickly the desired results. The existing committees both by virtue of their composition and by virtue of their structure can hardly be expected to fulfill the function which is required. This is fully borne out by the experience gathered during the last 10 months. . . . [22]

In spite of the increasing favourable reaction of the Government representatives little headway was made during the past ten months, either towards organising and financing the necessary experiments or towards securing an adequate supply of uranium ore for the future from the Belgian Kongo, which is the most important source of highgrade uranium ore.

At present it may be assumed that the experiments which will be carried out at Columbia University will find financial support through Dr. Bush's committee, but if each necessary step requires ten months of deliberation then obviously it will not be possible to carry out this development efficiently. Since April of this year it has been known that work on uranium is proceeding in Germany in great secrecy and on a very large scale in two of the Kaiser Wilhelm Institutes under the auspices of the German Government. This is precisely what was predicted by Prof. Einstein in his above mentioned letter of September last year.

5. Suggestions

In order to insure that the task before us is carried out with the efficiency of a going enterprise it is proposed that it be entrusted to a non-profit organisation which is formed for the purpose. The scientists responsible for devising this project and who are familiar with the various aspects of this complex material ought to be included in the executive and ought to be in direct touch with Dr. Bush and such other government representatives as are interested in the details of the project. Large scale experiments ought to be carried out as a joint enterprise of this organisation and Columbia University and such other universities as may be willing to collaborate and to put up required space and other facilities. It would be the responsibility of this organisation:

---

[22]We omit three paragraphs summarizing the history of the work.

1. To see to it that all necessary experiments be carried out at one place or another and

2. To see to it that all necessary materials be available for such experiments in the required quantity and quality.

3. To find out in what form if any collaboration with industrial organisations such as for instance Westinghouse, General Electric and Dupont is desirable and possible, and if desirable to establish such collaboration.

4. To maintain contact with the Canadian and U.S. producers of uranium and to stimulate if necessary an expansion of the production.

5. To advise the government in general and the Secretary of the Navy in particular of the developments and to prepare the ground for gradually transferring the experience acquired to the Navy at the appropriate time.

It may be mentioned that this form of organisation has been repeatedly discussed at previous meetings and an opinion strongly in favour of it has been expressed by Prof. Einstein in a letter addressed to Dr. Briggs, a copy of which is enclosed.

It is proposed that the seat of the organisation be in New York City, that the board of trustees include the names of Prof. Pegram, Prof. Urey, Prof. Lawrence, Prof. Du Bridge,[23] Prof. Wigner, Dr. Sachs and if government employees be included the names of Dr. Briggs and Admiral Bowen. It is proposed that Prof. Pegram be chairman of the board and Dr. Sachs act as treasurer.

It is proposed that the executive be composed of Dr. Pegram, Dr. Urey, Dr. Fermi, Dr. Szilard and Dr. Sachs, all of New York City and that a committee of scientists be responsible for supervising all the work, which committee includes the names of H. C. Urey, M. A. Tuve, G. Breit, G. B. Pegram, E. Fermi, L. Szilard, E. P. Wigner, [and] E. Teller.

It is proposed that a fund of $20,000 be put at the disposal of the trustees of such an organisation.[24]

Szilard was at last given a paid position under government contract, but he was seriously concerned that his position and the organization of the project be such as to ensure its success. We do not know whether the following was sent.

Document 90

[about October 1940]

Dear Professor Pegram,

For reasons which I shall attempt to explain further below I think it might be desirable

---

[23]Lee A. Dubridge, University of Rochester, director of the Radiation Laboratory, MIT.
[24]Sachs' revision of this memorandum, October 1940, followed this general line and made the same recommendations, except that the sum mentioned was $140,000, the amount which Bush's committee recommended be allocated for such work. But the government preferred working through contracts with universities and existing industrial corporations.

that I should put on record certain inventions which I have made within the last two years before I actually start to work under the contract with Columbia University on experiments for which the funds are being provided by the Government.

If you see no particular objection to the following I would propose to choose the form of patent applications which I would file before starting the projected experiments at Columbia. I would submit copies of these patent applications through you to Dr. Briggs and state in an accompanying letter that I would be glad to assign these patent applications now or at any time later during my work under the proposed contract with Columbia to such agency of the Government as may be suggested by you or the chairman of the uranium committee, without expecting any financial compensation from the Government, but with the understanding that these patent applications would be re-assigned to me at my request if at some future date the Government abandons the project in support of which it has now granted the sum of $40,000. In that case I would first try to obtain the support of the Rockefeller or Carnegie Foundation for the continuance of the work, and if that fails I would ask for a re-assignment of these patents in order to use them for inducing some industrial corporation in the United States or in Canada to support the project further until its successful conclusion. If you or Dr. Briggs should feel that in this case the Government ought to have the use of these patents even after their re-assignment to me free of royalty, I should be glad to give an undertaking to that effect.

I shall, of course, see to it that no patent is issued as long as the Government feels that secrecy is desirable, and such secrecy could be assured if Dr. Briggs wrote to the Commissioner of Patents, asking him to take the steps to this effect, which are prescribed by the law. This can be done even if the patent is not, or not yet, assigned to any agency of the Government.

If Dr. Briggs does not see any objection I might apply for a secret patent in Canada and in England, which I would assign without financial compensation to the respective governments, but I shall, of course, refrain from doing this unless I hear from Dr. Briggs that there would be no objection on his part or on the part of any other Government agency which is interested in the matter.

Of course, I should like to remain free to withdraw this patent application at a later date as long as it has not been assigned to the Government.

My reasons for the above proposal are the following: In July last year I formed the conviction that it will be possible to maintain a chain reaction with uranium under conditions which are satisfactory from an engineering point of view, that is under conditions in which it is possible to have the reaction take place at high temperatures and accordingly at a high rate on an industrial scale. One of the primary objects is the construction of an engine which allows it to utilize the heat liberated in the chain reaction for purposes of power production, primarily perhaps for locomotion. A number of problems arises in this connection, in particular problems of heat transfer and regulation, which present considerable difficulties but which seem capable of a satisfactory solution.

Having contacted a number of my colleagues in July and August last year, a consensus of opinion developed that an attempt should be made to enlist the support of the Government for carrying out a certain project until its successful conclusion. It may be half a million to a million dollars will have to be spent before the first engine of this type can be made to work, at least to the extent as to demonstrate the liberation of nuclear energy on an industrial scale.

I now understand that a certain sum has been granted by the Government for the support of experiments which will be carried out at Columbia University under the immediate direction of Fermi and myself. I need not emphasize how very happy I am that we shall thus be put into the position of starting to work along the lines which [we] discussed at various times with various representatives of the Government. While we have at present no assurance that further support will be forthcoming we have to base the work on the hope that this project will be supported by the Government until its successful conclusion, i.e., until the construction of the first engine working with nuclear energy is completed.

I personally have repeatedly expressed great confidence that it will be possible to have a nuclear chain reaction in unseparated uranium, and in accordance with this conviction I propose to dedicate the next five or ten years to this task.

I realize, however, that it is difficult to convey this faith in the ultimate success of the work with unseparated uranium to others, and that others take now, or will take after some of the set-backs which we may suffer at one time or the other in the future, a considerably less optimistic view. For this reason we have to envisage the possibility of not being able to get adequate Government support for certain rather expensive experiments which may become necessary. In that case I should be in favor of applying for support to the Rockefeller or Carnegie Foundation rather than to transfer this work within the framework of some industrial corporation. If such support is not obtainable then I would, as a last resort, be in favor of using the proposed patent application as an inducement to some industrial corporation in this country or in Canada for taking up this development, thereby giving them the hope that they might recover the expenses of development through profits which they might make in the future.

I fully understand that any development arising out of the work which we shall carry out under the contract between Columbia University and the Government will become the property of the Government and that no patents may be taken out by those who are engaged in this work.

Leo Szilard

Chapter V
Somehow we did not seem to get the things done which needed to be done.

My main concern was to get uranium (if possible in the form of metal) of sufficient purity and to get graphite of sufficient purity to make a valid experiment. What we mainly wanted to do was to test directly, by measurements on a pile composed of graphite and a lattice of uranium spheres, whether or not a self-sustaining chain reaction could be expected to occur if the pile were made sufficiently large. We did not have funds to purchase materials in any appreciable quantity, but [Briggs?] promised that such funds would be made available through the National Bureau of Standards, which was supposed to purchase these materials for us.

The trouble was that these materials could not be obtained in sufficient purity commercially. Having to negotiate for them through the Bureau of Standards became a major bottleneck of progress. Direct contact with manufacturers of materials is very important if no finished product is commercially available, because only through private conversations can you discover how the quality of the material might be improved. One important fact came out of a casual conversation with representatives of the National Carbon Company. Fermi and I had lunch with two men from the National Carbon Company, from whom we expected to buy some graphite. The graphite seemed to be fairly pure, and the total impurity would have been dangerous only if it had contained some element that very strongly absorbed neutrons. When we had our luncheon I said half-jokingly to one of these men, "You wouldn't put boron into your graphite, or would you?"[1] The two men looked at each other and there was an embarrassed silence. "As a matter of fact," said one of them, "samples of graphite which come from one of our factories contain boron, because it so happens that we manufacture in that factory graphite electrodes for electric arcs, into which boron is customarily put." Had we negotiated as we were supposed to do with these men—through the National Bureau of Standards—we would never have discovered this important fact. (*Docs. 91–92*)

We had worse luck with uranium. We were given the specifications of the uranium oxide which was supposed to be delivered to us, and the uranium seemed to be pure enough on the basis of those specifications. But then on a visit to the factory which made uranium metal out of the uranium oxide for our experiments we discovered another list of impurities, which differed from ours and which was much worse. This was a purely accidental discovery. It led us to reexamine the uranium which was delivered to us, and it turned out that our uranium was equally impure. When I looked into the process of how the uranium was purified, I was struck by the fact that an important group of elements which were strong neutron absorbers . . . were never removed from the finished product. But when I discussed with the National Bureau of

---

[1] Very small traces of boron are sufficient to render graphite useless as a moderator.

Standards whether we shouldn't change the procedure of purification, I was told that the process which would improve the quality would take a long time to prepare, and since we were in a hurry to get the chain reaction going, the Bureau of Standards was not willing to advocate a change in the chemical purification.

Because all these troubles were besetting us I got more and more impatient during the first half of 1941. Somehow we did not seem to be able to get the things done which we knew needed to be done. (*Docs. 93–95*) (*Fig. 5*)

During this early period I was also haunted by the fear that it might be possible to detonate uranium metal by fast neutrons, if a sufficiently large quantity of this metal were assembled. Whether or not this would be possible depended on the following thing: the bulk of natural uranium is uranium-238, and it fissions only if it is hit by fast neutrons. In this fission it emits fast neutrons, and whether or not a chain reaction can be maintained depends on how fast the neutrons emitted from fission are slowed down so that they might lose their effectiveness [in splitting] further uranium. Dr. [?] and I therefore pursued a side-line investigation to determine how fast uranium metal slows down fast neutrons. We did not stop this line of investigation until we were satisfied that uranium metal could not be used to make a bomb. [*10*]

While up to this point we had suffered from the lack of official recognition, during this period we were suffering from having official recognition. [*2*] Things would have dragged on in a most unsatisfactory way had not the British recognized that it is possible to separate a sufficient quantity of uranium-235 to make atomic bombs. Anybody could have recognized this fact. We knew two things: how much uranium-235 could be separated with a reasonable industrial effort, and how much uranium-235 it took to make a bomb. Urey, at Columbia University, and the Office of Naval Research worked on the separation of the isotope uranium-235, while Fermi and I worked on the nuclear properties of uranium. It so happens that I actually measured the cross-section of uranium-235 for medium-velocity neutrons in the first half of 1939. From this I could have computed how much uranium-235 it takes to make a bomb. The amount seemed fairly large, and I did not know that it was possible to separate such quantities of uranium-235. Urey's contract specified that he was not supposed to discuss his results with Fermi and me, who were not cleared. Therefore we were not able to put two and two together and come out with a simple statement that bombs could be made out of reasonable quantities of uranium-235. (*Doc. 98, Part IV*)

In Britain there were a number of German refugees such as Simon, Peierls, and Frisch, who at the beginning of the war were not permitted to work on anything of military significance, and therefore took to working on uranium. Simon was interested in the separation of uranium-235; Frisch and Peierls were interested in nuclear properties.[2] Nothing prevented them from talking to each other. They put two and two together and they informed the British government of the possibility of making

---

[2]Physicists Francis Simon at Oxford, and Rudolf Peierls and Otto Frisch at the University of Birmingham. Actually it was Frisch who was particularly interested in isotope separation; Peierls had a special interest in the critical mass problem. Margaret Gowing, *Britain and Atomic Energy, 1939–1945* (N.Y.: St. Martin's Press, 1964), pp. 40ff.

SECRET

CLASSIFICATION CANCELLED
Date 4/21/56
For The Atomic Energy Commission
C. L. Marshall/ncr
Director, Division of Classification

MEMORANDUM

L. Szilard

September 19, 1942

Introduction.

These lines are primarily addressed to those with whom I have shared for years the burden of knowing that it is within our power to construct atomic bombs. What the existence of these bombs will mean we all know. It will bring disaster upon the world if the Germans will be ready before we are. It may bring disaster upon the world even if we anticipate them and win the war, but lose the peace that will follow.

The requirements of keeping our work secret makes it impossible gradually to prepare the country for the role which it will have to assume immediately after the war is over. ~~If it wants to~~ have peace in a world where a lone airplane can drop ~~a~~ bomb over a city like Chicago. Not one single house may be left standing and the radioactive substance scattered by the bomb may make the area uninhabitable for some time to come. Some way will have to be found to ~~prepare for~~ adequate action after the war, those whom the constitution has entrusted with determining the policy of this country.

Most of you may feel that of more immediate concern to us is the fact that the perhaps most promising branch of this work ~~known,~~ which is pursued at Chicago, is not developing as rapidly as it ought to.

What is the trouble with our project?

Roughly speaking there are two kinds of troubles which frustrate our work. Troubles which originate at the Washington end of our organization, and troubles which originate at the Chicago end of our organization.

The unsatisfactory state of our metal supply, nine months after the reorganization which lead to a merger of the Chicago and the Columbia projects,

This material contains information affecting the national defense of the United States within the meaning of the espionage laws, Title 18, U.S.C., Secs. 793 and 794, the transmission or revelation of which in any manner to an unauthorized person is prohibited by law.

SECRET

Figure 5
This is the first page of the first version of a Memorandum "What is Wrong with Us" with corrections in Szilard's handwriting.

uranium-235 bombs with quantities of material that were industrially available. The British government informed the American government. So for the first time our attention was directed to the problem of making atomic bombs rather than merely to the problem of making a chain reaction . . . For the first time the government realized that our project was important.

Oliphant[3] came over from England and attended a meeting of the Uranium Committee which neither Fermi nor I was permitted to attend. He realized that something was very wrong and that the work on uranium was not being pushed in was an effective way. [10] He travelled across this continent from the Atlantic to the Pacific and disregarding international etiquette told all those who were willing to listen what he thought of us. Considerations other than those of military security prevent me from revealing the exact expressions which he used. If Congress knew the true history of the atomic energy project, I have no doubt but that it would create a special medal to be given to meddling foreigners for distinguished services, and Dr. Oliphant would be the first to receive one. [8]

He discussed his concern with E. O. Lawrence, who in turn approached Compton,[4] and as a result of this agitation it was decided to reorganize the project. A. H. Compton was supposed to be in charge of setting up a chain reaction with a view to producing plutonium. Mr. Urey was supposed to be put in charge of separating uranium-235 by the diffusion method, and Lawrence was supposed to be in charge of separating uranium-235 by [the electromagnetic] method. Actually reorganization took place around . . . January 1942. At that time the project from Columbia University was moved to Chicago, and all of the grant funds were put at the disposal of the [Chicago] project. However, even now the authority to purchase materials was not given to the project.

While negotiations for materials formerly had to go through the Bureau of Standards, now negotiations for the purchase of materials had to go through [Eger V.] Murphree of Standard Oil of New Jersey. This division of authority hampered us through the first half of 1942. In spite of this, somehow A. H. Compton, director of the project, managed to make arrangements for obtaining uranium purified in the right way so as to [free?] uranium from the neutron absorbing substances which the older way of preparing it did not remove. As purer-grade uranium was obtained and pure uranium metal began to come in, it became clear that a self-sustaining chain reaction would be achieved. This much was clear to most of us, including A. H. Compton, in the spring of 1942. And on December 2, 1942, the chain reaction was actually started at Stagg Field on the campus of the university. [10] There was a crowd there and then Fermi and I stayed there alone. I shook hands with Fermi and I said I thought this day would go down as a black day in the history of mankind. [11]

Fermi is a genuine scientist. There is no greater praise I could bestown on anyone. The struggles of our times do not affect him very much, and he is no fighter. Some

---

[3]Marcus Oliphant, physicist at the University of Birmingham.
[4]Arthur Holly Compton, chairman of the physics department, University of Chicago.

people hold this against him; they have no right to do so in my opinion. During the war Fermi did whatever was expected of him. He gave the best advice he could and never lost his sense of proportions. He, like all of us, wanted us to win the war, but somewhere deep in his heart he knew that it did not really matter whether we won or lost, and that a hundred years from now people would not find it easy to understand what it was we all had gotten so excited about.

During the war in the uranium project we had often to fight against foolish decisions which were about to be taken and which might have lost us the war—or at least so many of us thought. Often we got mad; Fermi got mad too. "It makes me mad," he said on one occasion, "if only on the ground of aesthetic considerations." On matters scientific or technical there was rarely any disagreement between Fermi and myself or the rest of us whose opinions counted in those early days. Nor was there much difference in our attitude towards those who "directed" our work from Washington. "This one is as wrong as wrong can be," said Fermi on one occasion. Mostly we spoke of them as "they." "If we brought the bomb to them all ready-made on a silver platter, there would still be a fifty-fifty chance that they would mess it up," he said on another occasion. All this was before the Army took over.[5] Fermi has a very good way of making a point without using many words. "A general is a man who takes chances," he said. "Mostly he takes a fifty-fifty chance; if he happens to win three times in succession he is considered a great general."

But Fermi thought we took too great chances in the project; on this we thoroughly disagreed. Even by the middle of 1942 Fermi thought that our work had no bearing on the war and that those who thought otherwise were "sadly mistaken." Whether he did not wish that there should be a bomb or whether he did not make sufficient allowances for possibilities which had not yet become manifest I do not know, though as a rule whenever we disagreed the source of the disagreement lay outside the intellectual sphere. [5]

As soon as it became clear that the chain reaction would succeed, my attention and also the attention of others turned toward the problem of having an effective cooling system and of solving the technological and engineering problems belonging to these cooling systems, so that a reactor with high power output could be constructed and sufficient quantities of plutonium could be manufactured.[6] There was a feeling in the project that the cooling of a reactor was not a problem for physicists to worry about, that this was an engineering problem and should be entrusted to engineers. There was an engineering group set up in the project, which set up an advisory committee having eight members [May, 1942]. I was one of the members but E. P. Wigner was not put

---

[5]The U.S. Army took over the fission project, naming it the Manhattan Engineer District, in September, 1942.
[6]Production of plutonium in a reactor is proportional to the rate of the chain reaction, but the power—heat—generated is also proportional to this rate. Thus at a high rate of plutonium production the reactor will melt unless it is cooled; the limit on production is set by the efficiency of the cooling system. Once the problem of purity of materials was in hand, Szilard and others worried chiefly about cooling.

of the committee. Since it was clear that Wigner thought more about engineering problems, taking due regard to the physics involved, than anyone had, I tried to correct this omission. But I did not succeed, and the engineering group took a position that they did not want to enlarge the committee, for it would be unwieldy. [*10*]

I met Wigner for the first time in 1921 in Berlin. He studied chemical engineering at that time and was about to graduate. I majored in physics and merely dabbled in chemistry, but through some freak we were thrown together in the same laboratory course. He was much more advanced of course than I was, and though he was much younger than I was, I used to turn to him whenever my analysis did not seem to come out all right, and very often it did not. This set the pattern of our relationship. . . .

Wigner was the conscience of the [Manhattan] Project from its early beginning to its very end. His great courtesy is proverbial and there are many stories circulating about him with his politeness as the central theme. Sometimes he is misunderstood, for if he says to a young physicist who explains to him his experiments or theories, "This is very interesting," he might mean the same thing as I do if I say, "This sounds like God Damn nonsense to me." Wigner as the story will show is a fighter, and [Henry] Smyth says, "He is stubborn like a mule." A mule with Wigner's brain would have a right to be stubborn. [*5*]

This engineering group decided to adopt a cooling system such as a car cooling system [which I will call] cooling system number one,[7] and to develop the process design along this line. Neither I nor Dr. Wigner thought that this . . . approach of the cooling system number one was acceptable. Wigner therefore tried to get an engineer attached to his group of physicists in order to work out an alternative system which I shall call system number two.[8] It took a long time before the engineering group agreed that he should have an engineer, but finally he was given an engineer, and they walked away, quietly trying to develop with one engineer and a number of physicists what we might call a cooling system number two.

When, at the end of 1942, the Du Pont company took over the construction of a plant for the manufacture of plutonium,[9] the official recommendation of the project was to adopt the cooling system number one advocated by the engineering group. After the Du Pont company had a few weeks' opportunity to study the system, Dr. Wigner presented to them a process designed for system number two, which he had worked out with one engineer.[10] The Du Pont company decided to use system number two rather than system number one which was unworkable. (*Doc. 98*)

Physicists in the project were unhappy about the way cooperation with the Du Pont company was set up. The Du Pont company had very good engineers, but they did not

---

[7]The Engineering Council at first adopted a reactor cooled with circulating helium gas. Szilard was meanwhile developing a speculative liquid bismuth cooling system. Hewlett and Anderson, *The New World*, 177ff; *Scientific Papers*, pp. 359–368.
[8]Water-cooled.
[9]In December Du Pont agreed to build a plutonium-producing reactor facility, later located at Hanford, Wash.
[10]Wigner and Gale Young, January 9, 1943.

have the required knowledge of nuclear physics. They were supposed to draw up the plans, and the [Chicago] project was given the right to object to any given solution which the Du Pont company might put forward. Clearly this was a very peculiar way of arriving at a design, and for a long time most physicists on the project did not believe that the Du Pont company would be able to produce a workable design on the basis of this trial and error procedure. . . .

The disagreement about the cooling system to be used was in the form of a fight between the physicists and the engineers. The issue was, should the physicists be permitted to make their own designs, or should all designing be concentrated in the engineering group and the physicists merely act in an advisory capacity? This fight reached its end, of course, when the Du Pont company took over the construction of the plant, for the responsibility was then clearly assigned to the Du Pont company. The fight gradually ended with the temporary victory of the engineers. Dr. Compton instructed the physicists to cease working on the process designs and to act as consultants to the engineering group.

When the engineers came over to ask Dr. Wigner for his cooperation, Dr. Wigner asked, "What do you want me to do?" "Well," they said, "all we want you to do is to answer our questions." "Oh," said Wigner, "if you know what questions to ask, you will find the answer to any question which you might ask and which I can answer in my files. All I have to do, then, is give you the key to my files, which I shall be very glad to do." Obviously, in order to get the right answers you must know what the right questions are. This kind of cooperation would have led us nowhere had we in fact adopted it. [*10*]

It is impossible truthfully to record the early story of the atomic bomb without hurting the feelings of some of those who were involved. I am sorry about this. If it were not for my addiction [to the truth] I could temper the truth with kindness, at least more so than it is possible for me to do in the circumstances. I do not wish to say of course that other accounts of the "early history," those written in the past or those which may be written in the future, are less truthful because they are different from mine. In 1943 Hans Bethe from Cornell visited in Chicago and we discussed the work conducted there under the Manhattan Project in which I was involved. The things that were done and even more the things that were left undone disturbed me very much particularly because I thought (quite wrongly as we now know) that the Germans were ahead of us. "Bethe," I said, "I am going to write down all that is going on these days in the project. I am just going to write down the facts—not for anyone to read, just for God." "Don't you think God knows the facts?" Bethe asked. "Maybe he does," I said—"but not *this* version of the facts." [*5*]

Documents from December 1940 through February 1944

Many letters, such as the following, show Szilard's constant work on the problem of acquiring pure materials.

Document 91

c/o Department of Physics
Columbia University
New York City
December 16, 1940

Mr. V. C. Hamister
Research Laboratories
National Carbon Company
Edgewater Works
Cleveland, Ohio

Dear Mr. Hamister:

I am writing to you to tell you how much Professor Fermi and I enjoyed your visit. We feel that these conversations might prove very fruitful, and hope that there will be soon an opportunity to continue them.

I have asked Mr. MacPherson whether he could possibly send us the ash of half of a graphite brick. We feel that if we test the ash and in addition have a spectroscopic analysis, we can be fairly sure of not missing those impurities which interest us.

With best wishes,

Yours very sincerely,
Leo Szilard

Document 92

February 7, 1941

Mr. H. D. Batchelor, Director of Research
National Carbon Company, Inc.
Edgewater Works
Cleveland, Ohio

Dear Mr. Batchelor:

Many thanks for your kind letter of January 31. We appreciate very much the attention given to this matter by your Research Laboratory and investigations conducted by Messrs. Hamister and MacPherson, and regret to hear that you are not in a position to supply graphite bricks free of boron to meet certain specifications of ours.

We should be very much interested to learn though the boron content of the best graphite which you are able to supply. For certain uses of graphite, we would be able to tolerate more boron than for other uses, although we are interested in every case in keeping the boron content as low as possible. Perhaps your graphite could be used at least for some of our work.

Very truly yours
Leo Szilard

While continuing scientific work and exploration of raw material sources, Szilard never ceased to criticize and seek to improve the structure of the fission project. His anxiety and his conflicts with his administrative superiors were greatly sharpened by two factors: a necessity for secrecy, which promoted confusion and distrust, and fear that Germany would get nuclear weapons before the Americans did, which made every decision seem a matter of life or death.

Document 93

May 26, 1942

Dr. V. Bush
Office of Scientific Research and Development
1530 P Street, N. W.
Washington, D. C.

Dear Dr. Bush:

I am taking this step of writing to you because I am concerned about the slowness of the work on unseparated uranium. In the past those who had originated this work did not ask to be consulted on matters of organization which had vitally affected their work. This, I now believe, was a mistake on our part. If we had presented to you our views on such matters perhaps we might have been able to explain to you our difficulties and you might have been able to remove them.

At present the main source of our troubles seems to lie in a division of authority along the wrong lines. This was the cause of most of our difficulties from the start and the net result was as follows:

When we started to work under contract with the NDRC in November 1940 we had a simple task for which we were well prepared through the spade-work which had been done ahead of time. All we had to do was to pile up about 40 tons of graphite with 10 tons of uranium oxide and to perform a measurement which takes about one week. *These materials required were available in sufficient purity at the outset of our work.* We could have procured them and completed the experiment at any time within four months after the allocation of the funds. Instead of four months it took us from November, 1940 to May, 1942, i.e., 18 months to perform this task. I should be very glad to give you a detailed statement of these facts.

The reorganization which you undertook last Fall when you asked Compton, Lawrence, Murphree and Urey to take charge of different divisions of the work was an improvement but it created again a division of authority.

If our future task were as simple as the task which we have just completed we might muddle through with our present organization. Unfortunately this is not the case and our new task is of great complexity. Almost all of the knowledge and ability which it requires is represented in Dr. Compton's group, but with the present division of authority between Compton and Murphree neither of the two groups can function properly.

If you should find it possible to go over these matters with me, I would like to describe to you in just what way this division of authority is blocking the path to a successful conclusion of our work. It would also be necessary to discuss other factors which affect the speed of our work and endanger its success. I can come to Washington if you will let me know a few days ahead of time when you can see me. Your message would reach me this week, c/o Metallurgical Laboratory, University of Chicago.

We knew in August, 1939, how to make a power plant with graphite and uranium. By June, 1940, we knew how to make "copper"[1] and bombs sufficiently light to be carried by airplane. Some of our recommendations were embodied in a memorandum dated August 15, 1939, which was submitted to the President, and a memorandum dated October 21, 1939, which was submitted to Dr. Briggs. I wonder whether, if you read the enclosed copies, you might not think that the war would be over by now if those recommendations had been acted upon. Our recommendations concerning the form of organization best suited for this work were embodied in a letter written by Professor Einstein to Dr. Briggs in April, 1940.[2] The experience of the last eighteen months leads me to think that we would be able to move much faster if some such form of organization were adopted.

In 1939 the Government of the United States was given a unique opportunity by Providence; this opportunity was lost. Nobody can tell now whether we shall be ready before German bombs wipe out American cities. Such scanty information as we have about work in Germany is not reassuring and all one can say with certainty is that we could move at least twice as fast if our difficulties were eliminated.

Yours very truly,
Leo Szilard

[1] Plutonium.
[2] See *Doc. 79.*

Chapter V: Documents from December 1940 through February 1944

Document 94

Vannevar Bush, Director
Office for Emergency Management
Office of Scientific Research
and Development
1530 P Street NW.
Washington, D.C.
June 1, 1942

Dr. Leo Szilard
Columbia University
New York, N.Y.

Dear Dr. Szilard:

Your letter of May twenty-sixth has resulted in a conference between Dr. Conant and myself. I think I recognize some of the difficulties of the present organizational scheme, and I consider it as a transitional form only. More definite plans are being studied at the present time and I think the next steps will be in a direction to avoid some of the difficulties that you mention. Nevertheless, I would like very much to discuss this subject with you. In view of the past history which you review, some of which, as you know, occurred before I had any connection with the matter, I am accumulating some data from the files in order to be fully informed. When I have this in my hands I would much like to have you elaborate on your letter and do so in direct conference with me, and I trust that you can visit me for this purpose. If you will let me know when this will best fit in with your other plans, and drop me a note in that regard, I will then make an appointment just as promptly as I can fit it into my somewhat crowded schedule.

Very truly yours,
V. Bush
Director

The following memorandum reviews some history in detail in order to criticize the Project's organization. Szilard also ranges beyond to general questions of the future of nuclear weapons and the responsibilities of scientists.

Document 95

WHAT IS WRONG WITH US?
L. Szilard
September 21, 1942

Introduction

These lines are primarily addressed to those with whom I have shared for years the knowledge that it is within our power to construct atomic bombs. What the existence of these bombs will mean we all know. It will bring disaster upon the world even if we anticipate them and win the war, but lose the peace that will follow.

We cannot have peace in a world in which various sovereign nations have atomic bombs in the possession of their armies and any of these armies could win a war within twenty-four hours after it starts one. One has to visualize a world in which a lone airplane could appear over a big city like Chicago, drop his bomb, and thereby destroy the city in a single flash. Not one house may be left standing and the radioactive substances scattered by the bomb may make the area uninhabitable for some time to come.

It will be for those whom the constitution has entrusted with determining the policy of this country to take determined action near the end of the war in order to safeguard us from such a "peace". They will have to be prepared for this task in order to be able to fulfill it and some way will have to be found to do this.

Perhaps it would be well if we devoted more thought to the ultimate political necessities which will arise out of our present work. You may feel, however, that it is of more immediate concern to us that the work which is pursued at Chicago is not progressing as rapidly as it should.

Roughly speaking, there are two kinds of troubles which frustrate our work. Troubles which originate at [the] Washington end of our organization, and troubles which originate at the Chicago end of our organization.

The unsatisfactory state of our metal supply, nine months after the reorganization which led to a merger of the Chicago and the Columbia projects, is illustrative of the effect of the trouble which originates in Washington. But although this trouble originates outside of Chicago, I do not think that we ought to blame anybody but ourselves for this calamity. We knew from the outset that the division of authority between Murphree and Compton, with respect to processing our materials, would lead precisely to this sort of catastrophe. This was not only known privately to many members of our group, but it was openly stated by a number of us at an almost public joint meeting of Columbia and Chicago groups at Columbia in January. Even "outsiders" like D. P. Mitchell[3] and Smyth[4] saw the point and joined in the chorus of those who condemned the proposed arrangement as unworkable. There was not a single voice raised in favor of the proposal of dividing the authority for processing our materials between Compton and Murphree.

We may have to answer before history the question why we tolerated an arrangement which we knew could not work. It is not possible for us to shift the blame to Dr. Bush or Dr. Conant, who originally decided in favor of that arrangement without consulting us. They cannot devote their full time and attention to our problems, and this matter is of such complexity that nothing less than that will do. However intel-

---

[3]Dana P. Mitchell, physicist at Columbia.
[4]Henry D. Smyth, Chairman of the physics department, Princeton.

ligent a man may be, if he is not in direct contact with our problems, he is not able to foresee the consequences of decisions which affect the outcome of our work.

Under the circumstances it would have been our duty to tell Dr. Bush, in January, that the arrangement which he devised would not work. We might not have been able to persuade him that the arrangement was, objectively speaking, unworkable, but surely he would have known that no arrangement can be successful if the men who are supposed to make it work lack the faith that they *can* make it work.

We may be tempted, of course, to shift the blame on to Compton and say that it would have been for him to tell Dr. Bush that the scheme was unacceptable. This, I believe, would be unfair to Compton, and on this point I have to elaborate because it has an important bearing on the future, as well as on the past.

Our project is exceptionally rich in men who belong to the creative type and represent what may be called the artistic temperament. Compton is one of them, and there are quite a few others. To be sure, each of these men has his shortcomings and limitations. However, it would be a grave mistake to believe that because each of us has certain obvious weaknesses, the group as a whole is not fit to carry the full responsibility for its task. If we were properly organized, those shortcomings and weaknesses would not add up, but rather cancel out, within the group. If we were properly organized, there would be no task in physics, engineering or production, which we could not tackle and master, as long as each of us realizes the limitations of the others and sincerely tries to find out something about his own limitations.

Compton was put in charge of our project last fall by Dr. Bush, but if the question of leadership had been put up to our group, he would have been elected by unanimous vote. We may complain about his not taking a strong stand in Washington, but let us be clear on this point above all: We cannot eat our cake and have it; if we want a man who can run our project at Chicago and create here an atmosphere of friendship and loyalty, we cannot expect the same to go to Dr. Bush and threaten to resign unless he is given all the authority required for the success of our project, or to go over the head of Dr. Bush to the President, and ask the President for this authority.

This is amply borne out by past experience. Only rarely is it possible to make a forecast with a high degree of assurance and such a clear case as we had in January with respect to the division of responsibility between Compton and Murphree may not occur again. If Compton did not take action then, he cannot be expected to take action in any other case which is less clear cut as long as he has to ask for things for himself.

The situation might be different if Compton considered himself as our representative in Washington and asked in our name for whatever was necessary to make our project successful. He could then refuse to make a decision on any of the issues which affect our work until he had an opportunity fully to discuss the matter with us.

Viewed in this light, it ought to be clear to us that we, and we alone, are to be blamed for the frustration of our work which originates from the Washington end of our organization. We should have asked Compton to make our views clear, or if he should have preferred this, we should have made our stand clear in Washington ourselves. It is my personal conviction, however, that we shall not be able to make progress in this direction until we have put our own house in order.

What is Wrong at the Chicago End?

While the metal situation illustrates the trouble which originates in Washington, the fact that we have not developed a satisfactory cooling system and are not mentally prepared to do so in the near future, illustrates the trouble which originates at the Chicago end of our organization.

In one respect at least, the troubles which originate at Chicago are less serious since it is entirely within the authority of Compton to remedy the situation.

Stated in abstract form, the trouble at Chicago arises out of the fact that the work is organized along somewhat authoritative rather than democratic lines. There is a sprinkling of democratic spots here and there, but they do not form a coherent network which could be functional. This is partly due to a compartmentalization of information which is imposed on Compton from Washington. Since this is again trouble which does not originate at Chicago, we may pass it over here until such time as we have solved the "local" Chicago problems.

In order that we should be able to do so, it is necessary to realize that there are certain inherent difficulties which cannot be removed and must be met by skillful adjustment of our organization.

I believe that we ought to say at the outset that the breakdown of the Chicago organization is of our own making. Though certain trends followed by Compton which will have to be mentioned later do represent obstacles standing in the way of a well functioning organization, it was entirely within our power to compensate for these trends by making full use of the existing machinery. This we did not do and therefore we may say that the blame is ours, and ours only. At the most we can say in our defense that there were mitigating circumstances.

Compton likes to avoid violent clashes of opinions which are unavoidable within a committee small or large if vital issues are debated, on which strong feeling prevails. In accordance with this, he has a tendency to steer a meeting towards greatest harmony rather than toward vital decisions. He prefers to deal with the more important issues in private conversations, a method which is either time consuming, or leaves many members of the planning board uninformed. But even in private conversation with Compton I personally find it difficult to have an issue settled. Perhaps this is my fault. I am, as a rule, rather outspoken, and if I do not call a spade a spade I find it rather difficult to find a suitable name for it. It may be that in talking to Compton I am overplaying a delicate instrument. This is, by the way, an opportunity to apologize to all members of our group for my outspokenness and to ask them to consider it as one of the inevitable hardships of the war.

Just as the question of the metal supply illustrates the trouble at the Washington end, the question of the choice of a cooling system for the power unit illustrates the difficulties at the Chicago end.

We knew for a long time that there are three possible ways in which the power plant could be cooled: by helium, by water and by bismuth. The question whether these three systems should be designed simultaneously and the design be carried out possibly up to the blueprint stage on all three of them, or whether one of these systems should be worked out in preference to others, and which should be given preference, was never put up for decision to any of our committee. Even so these committees should have considered as one of their most important tasks to decide this question at an early date. No such decision was taken.

By May, it became apparent to everybody that k will go over 1.[5] An engineering group turned up somehow in our project and started to work, under the direction of Mr. Moore,[6] on plans for the helium cooled system. Nobody knew why the helium cooled system was given preference by this group, but it was the general impression that somehow, somewhere, somebody decided that we shall build a helium cooled system first and place orders for the heavy accessory equipment which the system required at an early date.

In the meantime, Wigner and his division became more and more convinced that a water cooled system could be built in a much shorter time, although they were not willing to say that they could foresee with certainty all questions of operational safety. The question whether Wigner would be willing to assume the responsibility for developing this system, if necessary, into the blueprint stage, was never put to him, and remains undecided up until the present day. Clearly, if Wigner were willing to assume this responsibility, an engineering staff would have to be added to his division in order to help them to produce usable plans for a water cooled unit. A proper balance between physicists and engineers in Wigner's division would require the addition of quite a number of engineers to this division.

It has been my personal opinion that it is not possible to judge the relative advantages and disadvantages of any of the three systems until they have been designed in detail, although it may not be necessary to carry the design into the blueprint stage. I further held the view that, particularly in the case of the water and bismuth cooled power units, certain simple technological tests will have to be made before the detailed designing of the power unit can be carried through with faith in its ultimate success. A thorough designing job can hardly be accomplished in the absence of such faith. I have therefore pressed for the establishment of a technological division which after a while came into existence.

I was put in charge of this division which had the task of looking after the technological problems involved in the three different cooling systems. For a time, it looked as if we might have three almost different groups, each comprising physicists and en-

[5]k is the "reproduction factor" for a reactor; if it is greater than 1, the reactor will be critical, i.e., working.
[6]Thomas V. Moore, chief engineer at the Metallurgical Laboratory.

gineers, to work simultaneously on designs for these different cooling systems. It is still my opinion that that would have been the right course of action. While I personally felt that I can contribute most towards developing the bismuth cooled system because I have more faith in this system than in the others, I am quite aware of the fact that this may be a purely personal preference, and I do not have any well founded opinion as to which of these three systems shall prove to be the most successful, or can be made to work fastest.

I, myself, was quite prepared to assume the responsibility for developing a design for the bismuth cooled plant. I felt, however, that this work could not be pursued with confidence unless the metallurgical problems which are involved were settled. As far as I could tell, Compton was in full agreement with this approach, and thanks to his vigorous help we succeeded in having Foote, a metallurgist, released from Cooper Union. Simultaneously, I contacted various organizations with a view of having a pump designed for liquid bismuth. Four tons of bismuth were ordered at a cost of $10,000.

Thus I made arrangement to divide my time between looking after the technological division and the designing of a bismuth cooled power unit. These two tasks are almost too much for one man and the only reason why I thought that I would be able to manage them was the fact that Creutz,[7] Foote and Marshall[8] can very well carry most of the burden of the technological division. Foote was supposed to look after the bismuth, Marshall after the recasting of uranium metal, and Creutz after the problems of the water cooled plant, and many other things as well. What became of all these plans?

About six weeks ago I was informed that "we are sending through Marshall and Creutz to Boston to fuse Alexander's metal at M.I.T." If Creutz had left, the whole technological division would have collapsed and the work of the theoretical division working on the water cooled power unit would have been frustrated, since Creutz was looking after the technology of the water cooled unit. I succeeded in substituting Foote for Creutz. I agreed to his departure because I assumed that it would be for two weeks at the most and besides I had reached the conclusion that it is no longer worth while to fight about individual issues.

It was not possible to have a date set for Marshall's and Foote's return in six weeks. All work on the bismuth cooled plant is stopped. While I may be wrong in laying such emphasis on the bismuth cooled plant, it is a fact that nobody else has looked into this matter, with the possible exception of the theoretical division, and consequently there is nobody in a position to express an opinion whether the damage brought about by Foote's and Marshall's continued absence is counterbalanced by the good they can do at M.I.T. So far nobody has made an effort to find men who could replace Foote and Marshall at M.I.T.

[7] Edward C. Creutz, physicist and metallurgist.
[8] John Marshall, Jr., physicist.

In the meantime, Dr. Moore's division carried the design of the helium plant into considerable detail. We were all under the impression that it has been decided to build a helium cooled plant and to place an order for heavy parts of the equipment at an early date. For this reason, while I had been studying this type of cooling at the earlier stages of our project, I did not continue to pay any attention to the questions connected with the helium cooling.

Suddenly, about a week ago, a few of us were informed that we are expected to express an opinion whether a helium cooled plant should or should not be built. I possibly could have formed such an opinion if I had followed the development of the helium cooled power plant for the past two or three months and I should certainly have been very glad to do so. In the circumstances, I do not believe that any opinion that I personally might now express on the subject would be worth anything, and I see no way how I could form a valid opinion within a short time.

Diagnosis

If I have to give a diagnosis in an abstract and therefore necessarily misleading form, I should say this: In the past, the men who were on the technical committee and the planning board did not have the feeling that they were responsible for the success of the project. There was no mechanism for reaching decisions and consequently decisions were reached in a haphazard way. Essential decisions were being omitted or were taken many months after they were due. As time went on, more and more of us began to emphasize that we do not want to be held responsible for what was or was not happening. There was more and more shrugging of shoulders among the group leaders and an increasing tendency of narrowing down their responsibility to this or that detail of the work with which they had been explicitly entrusted.

To all this came an increasing tendency of compartmentalization of information imposed upon Compton by the Washington end of our organization and inevitably our best men are losing the zest for their work. Recently, a major decision was taken by the Executive Committee which usually meets once a month in Washington.[9] This decision will vitally affect our project. It was the third major decision within a year which was taken without consulting our group, our planning board, or our technological committee. Finally, a stage has been reached where it becomes clear that we have to choose between two alternatives:

We may take the stand that the responsibility for the success of this work has been delegated by the President to Dr. Bush. It has been delegated by Dr. Bush to Dr. Conant. Dr. Conant delegates this responsibility (accompanied by only part of the necessary authority) to Compton. Compton delegates to each of us some particular

---

[9]The S-1 Executive Committee (Conant, chairman; Briggs, Compton, Lawrence, Murphree, Urey); on August 26 and September 13–14, 1942, they gave priority to electromagnetic separation of uranium-235.

task and we can lead a very pleasant life while we do our duty. We live in a pleasant part of a pleasant city, in the pleasant company of each other, and have in Dr. Comptom the most pleasant "boss" we could wish to have. There is every reason why we should be happy and since there is a war on, we are even willing to work overtime.

Alternatively, we may take the stand that those who have originated the work on this terrible weapon and those who have materially contributed to its development, have, before God and the World, the duty to see to it that it should be ready to be used at the proper time and in the proper way.

I believe that each of us has now to decide where he feels that his responsibility lies.

Document 96

November 25, 1942

[to] A. H. Compton
[from] L. Szilard

[re] Compartmentalization of Information and the Effect of Impurities of 49.[10]

We had repeatedly discussed, in an abstract way, the harmful effects of compartmentalization of information. In view of the present interest in the effect of impurities of 49 I should like to put on record for your information the following facts:

I realized this summer that the strong alpha emission of 49 would lead to a strong neutron emission from impurities which are expected to be present. Upon Teller's return from the summer conference in Berkeley, I went to see him and asked him whether this point had been considered and whether, in view of this fact, it would not be wise to put more emphasis on autocatalytic methods of explosion which I had discussed with him in the past.[11] Teller replied that he personally is placing considerable emphasis on the autocatalytic method, but that the group did not consider it as important. Teller, as you know, is an old friend of mine and I found him occasionally embarrassed when I tried to discuss with him things which he did not feel free to discuss with me. For this reason I interpreted his reply as meaning that the situation was well in hand, and that there was no need for us to discuss it. Consequently, I changed the subject of our coversation and did nothing further in the matter. Teller, on the other hand, as it now turns out, interpreted my changing of the subject as meaning that I was going to look into this question, and that it was not necessary for him to do anything about it.

I believe this illustrates well under what strain the scientists are being put, and how compartmentalization of information poisons the discussion, even in those fields which are not explicitly excluded from discussion.

---

[10]Code for plutonium.
[11]This was probably initiation of the explosion by neutrons normally produced in the plutonium, rather than by a special neutron source.

I do not believe that the question of impurities in 49 should be taken tragically, but it is a fact that if Teller and I could have talked freely, the question would have been raised three months ago, immediately following the summer conference in Berkeley.

cc: E. Teller

Szilard's concern with the postwar implications of the atomic bomb, already apparent in the first pages of Doc. 95, was connected with wartime use of the bomb as indicated in the last paragraph of the following letter:

Document 97

Metallurgical Laboratory
January 14, 1944

Dr. V. Bush
1530 P Street, N. W.
Washington, D.C.

Dear Dr. Bush:

I wish to thank you for your letter dated January 6th which I found on my return from a short trip. In the circumstances, it is perhaps best that I explain to you, as well as I can, the reasons for my having asked you for an interview. Then you can see whether or not such an interview could serve a useful purpose, and if you feel justified in taking time from other matters which may have a prior claim on your attention, naturally I shall be at your disposal.

It seems that the vast majority of the scientists who are working on the uranium problem within several special projects, and who are given enough information to be in a position to judge, believe that important branches of this work are crippled by circumstances which are not within the power of any one special project to remedy. Due to these circumstances promising branches of the work have been suppressed; other branches which are strongly pushed and given strong financial backing are handicapped, progress slowly, and even their ultimate success is in jeopardy.

A comparatively minor cause of regret, but one which is easily stated is the enormous expenditure in money, labor, and material devoted to the project in the North-west Pacific[12] which is universally looked upon as arising out of irrational considerations of safety and also exaggerated considerations of operational safety. Such considerations have led to placing individual power units ten miles from each other and consequently to an enormously extended area of manufacture which necessitated the building of 180 miles of roads and 160 miles of railroads.

I should make clear, however, that in anything I am saying in this letter I am leaving entirely out of consideration the opinions of Drs. Compton, Urey, Lawrence, and

---

[12]The Hanford, Washington, plutonium-production reactors.

Oppenheimer as well as those who act as their representatives since, due to their official position, some of them may not talk as freely as they otherwise might and also because I want to be sure not to commit an indiscretion with respect to any of them.

Apart from this group I do not know of a single man to whom I have spoken who did not express, in private conversation, his concern. Perhaps you will share my view that this fact alone is significant and indicative of some serious trouble of an objective nature. What makes this fact to me even more disturbing are the reasons which these men give for remaining silent and particularly for not endeavoring to put their views before you or Dr. Conant. It seems that very few are left who still believe that the situation can be remedied by approaching either of you. There are of course those who by now believe that the situation cannot be remedied by any method as long as the war is on and a few others who are simply not willing to stick their necks out now that powerful interests have become involved.

All in all I know by now of fifteen different persons who at one time or another felt so strongly about the situation that they intended to reach the President and all of them stood to lose something and had certainly nothing to gain by taking such an unusual step.

I wrote to you early in December shortly after a man, who had been during the summer months a member of the Chicago project, had some conversations in Washington and after I sought and obtained enough information about those conversations to fear that a certain amount of confusion might very well be the outcome.[13] I felt at this time that it is only fair that you should be acquainted with the existence of such widespread dissatisfaction. It appeared conceivable at that time that your opinion might officially be asked and it appeared therefore desirable that you be acquainted with the nature and cause of the dissatisfaction and be able to form an opinion to what extent this dissatisfaction is justified.

If you would allow me to do so, I could go over the history of the various special projects with you in order to let you see in what manner circumstances beyond their control have been affecting the progress of the work. If sufficient time were available, I think you could gain insight into the mechanism by which decisions come about and see that such decisions are often clearly recognized as mistakes at the time when they are made by those who are competent to judge, but that there is no mechanism by which their collective views would find expression or become a matter of record. The question is an intricate one and I know of no method for gaining insight into it other than the method of going in some detail over the history of the various decisions which were involved.

Since we cannot change the past obviously the main purpose of giving such an analysis is to learn in what manner the situation could be improved in the future and it is with this end in view that I thought the past ought to be scrutinized.

---

[13]See James B. Conant, *My Several Lives. Memoirs of a Social Inventor* (N.Y.: Harper & Row, 1970), p. 295; Hewlett and Anderson, *The New World*, p. 203.

One of the reasons I wrote to you rather than to Dr. Conant is the fact that you have been rather aloof from these matters and as far as I know have not taken an active part in the major decisions of the last eighteen months. Incidentally, as you may know, I myself have been rather aloof from all controversial decisions for well over a year[14] and so it may be that I shall escape the charge of having an axe to grind. However, just because Dr. Conant took an active part in a number of important decisions I would not feel at ease if I had to talk to you about them in his absence. Therefore, if an interview were arranged at all I should feel much happier if Dr. Conant could be asked whether he would care to participate.

From what I believe to be true, it would seem that the Berkeley project[15] is not suffering from those circumstances to which I referred above. Unfortunately, in the case of the Berkeley project a large expenditure in terms of money, skilled labor and power consumption is and will remain unavoidable since it arises out of the nature of the project rather than waste due to avoidable mistakes. More serious than the expenditure is the fact that the completion of the Berkeley project to full scale cannot in any case proceed faster than at a certain rate. Thus it will take a longer time for the Berkeley method to be in full scale production than it might take for some of the other methods if they were pursued under favorable conditions.

Finally, with your permission I may perhaps say why it appears to me (apart from the most obvious reason, and assuming that through good luck we shall end the war with Germany before the Germans have a chance to use this weapon upon us) so very important to do all that is in our power to have the final product in sufficient quantities at an early date. If peace is organized before it has penetrated the public's mind that the potentialities of atomic bombs are a reality, it will be impossible to have a peace that is based on reality. The people of this country clearly will not be willing to pay the price that, in the face of the existence of atomic bombs, is a prerequisite for a stable peace. Making some allowances for the further development of the atomic bomb in the next few years which we may take for granted will be made, this weapon will be so powerful that there can be no peace if it is simultaneously in the possession of any two powers unless these two powers are bound by an indissoluble political union. It would therefore be imperative rigidly to control all deposits, if necessary by force, and it will hardly be possible to get political action along that line unless high efficiency atomic bombs have actually been used in this war and the fact of their destructive power has deeply penetrated the mind of the public. This for me personally is perhaps the main reason for being distressed by what I see happening around me.[16]

Very sincerely yours,
Leo Szilard

[14] During 1943 Szilard was entangled in a dispute with the Army over patents and consequently was able to do little work at the Metallurgical Laboratory.
[15] Electromagnetic separation of uranium-235.
[16] In a four page reply, January 18, 1944, Bush stated that he felt the organization of the work was satisfactory and opinions could be freely expressed, but he was willing to discuss the problems further with Szilard. He did not refer to the issues raised in this final paragraph.

The following memorandum gives Szilard's views as he took them into a meeting he had with Bush in early 1944, and may have circulated in the Metallurgical Laboratory. It bears the heading "first rough draft." Of a total of 42 typed pages we give about 28, concentrating on Szilard's description of some events in the years 1940–1943.

Document 98

PROPOSED CONVERSATION WITH BUSH
February 28, 1944
L. Szilard

Part I

The Government sponsorship of the uranium work started over four years ago in October 1939 by the appointment of Dr. Briggs' committee by the President. At that time we were far ahead of other countries in knowledge, but this initial advantage we have presumably lost since that time. The work of the scientists was crippled from the start by a mistaken attitude on the part of the administrators toward the scientists upon whose discoveries and inventions all this work rests. Subsequently the same attitude was manifested toward most of the other *competent* scientists who joined this work later. A number of reorganizations took place and there were changes both in the agency and in the persons who were entrusted with the responsibility of administering this work but the attitude toward the scientists remained the same. There can be no substantial improvement unless there is a complete change of heart in this respect.

1. The worst consequence of this attitude and one from which many evils have derived is the fact that it is made impossible for the scientists (who are giving their *full* time and attention to various aspects of the uranium work) to have discussion with each other of the pertinent issues. It is therefore difficult for them to form a well-founded opinion and even if they individually arrive at definite opinions they are not able to put collective recommendations on record.

A direct consequence of this is that there can be no judgment on which the administration can base sound decisions on issues which often involve the expenditure of hundreds of millions of dollars. Most of the decisions made are based on false premises. . . .

2. From October 1939 to December 1941 [there was] inadequate financial support and complete frustration of the scientists in their attempts to make arrangements with industrial firms for the technological development which was considered by them a prerequisite of the industrial stage of the development. This affected the work of Dr. Urey on the separation of isotopes as well as the work of Dr. Fermi and myself on the chain reaction in unseparated uranium.

3. From January 1942 on there was adequate financial support for this work but

as far as the Chicago Laboratory is concerned, the Laboratory was not given the authority to make arrangements for obtaining materials which the scientists considered necessary for their work. This nearly wrecked the Chicago project. . . .

4. The free interchange of views between different groups working on uranium along related lines was prohibited beginning the fall of 1940. . . .

At present there are two methods for separating the uranium isotope in the industrial stage. One of these (K-25) is using the diffusion of uranium hexafluoride; the other (Y-12) is using a magnetic separation.[17] Both of these methods are clumsy methods. The first one may not even succeed. The second may have a good change of success, but it is very costly in time, materials, and skilled labor. It is more likely than not that if the competent men who are at present working for the Government on various aspects of the uranium problem had been given enough information to encourage them to think up better methods for separating the uranium isotope we would by now have much faster, simpler, and cheaper methods. This must not be interpreted as saying that the men who originated the present methods lack originality or ingenuity, but simply to say that there is no way of telling beforehand what man is likely to discover and invent a new method which will make the old methods obsolete. The only thing we can do in order to play safe is to encourage sufficiently large groups of scientists to think along those lines and to give them all the basic facts which they need to be encouraged to such activity. This was not done in the past and it is not being done at present. . . .

5. Another consequence of this situation is that certain large firms, particularly du Pont, are acting in a double capacity, both as contractors and as advisors of the War Department.

So far the work intimately connected with the chain reaction in unseparated uranium has been exclusively concentrated in the hands of du Pont. The engineering staff of du Pont is of course limited both in number and in men with sufficient theoretical background. Consequently they are not adequate to handle *every* problem that the scientists consider necessary to follow up. In two *important* instances of this type the scientists have attempted to obtain permission to cooperate with some other firm in order to push the development of certain alternative methods at least into the process design stage. In both cases this permission was refused.

There was an increasing tendency of regulating the work of scientists by means of directives, some of which are clearly based on false premises. While clearly a large fraction of the effort of the scientists has to be regulated by some sort of directives the success of the work is at stake if sweeping directives make it impossible to have 10 to 20% of the scientific staff pursue lines of development which the scientists themselves consider essential. . . .

As a result of all this the present state of the uranium work may be described as follows:

[17] Gaseous diffusion and electromagnetic plants, under construction at Oak Ridge, Tennessee. Both finally worked after much time and expense.

A. THE FIELD OF UNSEPARATED URANIUM

A water cooled production plant is being built at a cost of about $300 million by the du Pont Company. This plant is supposed to go into production on June 1, 1944. . . .[18] Mr. Wigner and all competent scientists in the Chicago Laboratory were . . . of the opinion that a water-cooled pilot plant of 10,000 or 20,000 kw should be built immediately. This was not done and three production units scheduled to work at 250,000 kw are thus being built with no other basis than experience based on a 1,000 kw *air cooled* plant. As it is there is hardly more than a fifty-fifty chance that the 250,000 kw production units will stand up to operation for a reasonable period of time. . . .[19]

B. SEPARATION OF ISOTOPES

(a) K-25, based on the diffusion of uranium hexafluoride, was transferred into the industrial stage and taken out from under Urey's supervision in 1942. This was done at a time when no pilot plant was in existence and it is said that at the time when the development was taken out of Urey's hands and placed in the hands of the Kellogg Company, there was not even an experimental unit in existence in which the diffusion method had actually been tested on the gas (uranium hexafluoride) to which it was to be applied. It is doubted that this system in its present industrial form has more than a fifty-fifty chance for success.

(b) In the fall of 1942 of the various methods for producing heavy water which were developed under Urey's supervision, one was chosen by du Pont and put into industrial production.[20] The scientists in Urey's laboratory were of the opinion that the process chosen by du Pont was much more costly and inefficient from the point of view of coal consumption than another method. . . . It was impossible, however, to obtain permission to collaborate with any firm other than du Pont in the development of this alternative method.

As a result of this our heavy water production does not exceed three tons a month and consumes 30,000 tons of coal per ton. The quantity of heavy water produced is not sufficient as a basis of adequate production of U^{233} or as a second line of defense for the production of plutonium. The production is expensive and we have no alternative method ready on which to fall back if a larger quantity of heavy water is needed or if economic conditions should compel us to discontinue the inefficient process used by du Pont.

(c) The electromagnetic method in its present form may very well be successful but

---

[18]These Hanford, Washington, reactors began production in early 1945.
[19]Szilard was concerned with the "Wigner effect"—degradation of a pile under intense neutron bombardment.
[20]Heavy water may be used instead of graphite in reactors that make plutonium (or the fissionable isotope uranium-233). For the Chicago group and heavy water see Weart, *Scientists in Power* (Cambridge, Mass.: Harvard University Press, in preparation).

it will require enormous expenditures of materials and skilled labor in order to supply the required quantities of the separated isotope. It is believed that on the present industrial scale the method will supply only a fraction of what is needed to win the war by this weapon. Even at this late stage, new inventions in the field of the electromagnetic method might bring about favorable changes in the situation but the competent physicists working on uranium in branches other than the special project of Lawrence are not supplied with information which is required in order to be encouraged to think up something new in this field.

Part II

The Exclusion Principle

A general principle seems to have consistently been applied to the conduct of the uranium work. This principle consists in not giving authority to act or authority officially to advise to men who have a direct knowledge of the intricate problems involved. It is preferred to give such authority to men who have no such direct knowledge and who know only what they are told about our problems. If one knows only what one is told, one does not know enough to be able to arrive at a well-balanced decision. . . .

MATERIAL PROCUREMENT IN 1942

At the end of 1941 a reorganization took place which was carried through by Dr. Bush and the Metallurgical Laboratory was set up in Chicago under the direction of Dr. Compton. When it became known that Dr. Compton was not given authority to make arrangements for procuring the materials which were needed for his project but that this authority was given to Dr. Murphree, it was clearly realized by the scientists that this division of authority might wreck the Chicago project. There was a unanimous opinion in this respect which was voiced at a meeting held under the chairmanship of Dr. Compton at Columbia University. However, this opinion was neither recorded nor communicated to Dr. Bush. Dr. Compton later on wrote and asked that such authority be given to him but it was not evident from his letter that his opinion was unanimously shared by all those who were competent to judge. The only way through which Dr. Bush gradually learned of this unsatisfactory situation was through individual conversations with Creutz, Wigner and Szilard, a rather irregular procedure which cannot as a rule be relied on in the conduct of such projects.

When it was finally recognized that this division of authority was harmful, attempts were made to remedy the situation. That these attempts were successful is due only to a series of very lucky circumstances which as a rule cannot be counted upon: One was personal friendship between Dr. Compton and the Mallinckrodt family through which Dr. Compton was able to arrange for a supply of chemically pure uranium oxide based on an ether purification of the uranium nitrate on an industrial scale (originally proposed by Dr. Creutz).[21] When the metal situation became desperate owing to the unsuccessful attempts of producing pure metal on a large scale by means of calcium reduction, I found the Brush Beryllium Company willing to make a few simple experiments and try to reduce uranium tetrafluoride by magnesium. I was not encouraged to make these arrangements, but through great luck the first few experiments already showed that pure uranium metal could be thus obtained by using commercial magnesium as a reducing agent. Since this result was obtained before the private arrangement with the Brush Beryllium Company was interfered with, it became possible for me to take a strong stand in favor of magnesium reduction. Finally, Dr. Spedding, a college professor, developed the magnesium reduction into a satisfactory industrial process and, by turning his laboratory into a factory, soon had the production up to two tons of metal per day and so saved the situation. . . . [22]

DESIGNING WORK IN 1942

By May 1942 Dr. Compton officially took the position that the chain reaction would work in the graphite uranium system. Within two days after the pure materials were available we had a chain reacting pile (December 2, 1942).

From May 1942 it was considered by many of the scientists at the Chicago Laboratory that our main task lies now in making a design for a chain reacting power unit which could dissipate several hundred thousand kw. Mainly due to outside influence, the official support of the laboratory went to an engineering group headed by Mr. Moore (recommended by Mr. Murphree) and this group decided to design a helium cooled power unit. The scientists were anxious that other methods of cooling should be developed at least into the process design stage. There were two such other cooling systems proposed—water cooling sponsored by Mr. Wigner, and cooling by bismuth or bismuth-lead alloy sponsored by me. Mr. Wigner found it difficult to get any support for developing the water cooled design but finally succeeded in having one engineer assigned to him. No such support was given to the development of the bismuth cooled system.

At first there was a committee called "engineering council" officially entrusted with supervising the engineering developments in Chicago. This council had weekly meetings, but this council did not invite Mr. Wigner to attend these meetings. I made repeated representations about this point but they were not successful, though I went to the length of going on record by writing a memo about it. . . .

DECISIONS BY EXECUTIVE COMMITTEE

This committee contained one man from each special project, i.e., the project leader, namely, Urey, Compton, and Lawrence. They were the only members of the com-

---

[21]The Mallinckrodt Chemical Company, St. Louis, approached in April, 1942, soon produced pure materials for experimental use.
[22]Frank H. Spedding, Iowa State College. See Hewlett and Anderson, pp. 293–294.

mittee who gave their full time and attention to the uranium work and the other members of the committee had no intimate contact with the scientific staff of the projects. This committee made recommendations on important issues.

It is difficult to say exactly what the recommendations of such a committee represent. Clearly Urey, Compton and Lawrence were the only members of the committee who knew enough of the details to be able to weigh most of the pertinent factors but even they, since they were exceedingly busy with the administration of their own projects, had only a limited knowledge of each others' projects. Moreover, none of them could push his own project in such a committee without putting himself into an awkward position and each of them had to lean backward if they had to give their opinion concerning one of the projects which competed with their own project.

This committee made, in the course of 1942, quite important decisions concerning which side-lines should be pushed and which side-lines should be dropped and the scientists had always an uneasy feeling concerning the wisdom of these decisions. Many believed that in this committee Dr. Conant's views always prevailed and so the committee's decisions were mostly regarded as Dr. Conant's decisions.

Dr. Conant at that time was far from being able to give his full time and attention to that matter and it ought to be generally recognized that no one who devotes only part of his time of these very complex questions can hope to have as good a chance to take the right decision as men who may be less capable but who devote all their attention to the uranium work. . . .

DECISIONS BY AD HOC APPOINTED COMMITTEES

Our first experience with an ad hoc appointed committee was as follows: In the fall of 1942 General Groves decided to try to place a contract with du Pont to build a power unit. Between November 2nd and 6th a large group of du Pont engineers visited the Laboratory. The helium cooled system was explained to them in great detail and half an hour was devoted to discussing with them the water cooled and metal bismuth cooled systems. This group of engineers came to the conclusion that the helium system was the best, next they placed the homogeneous heavy water system, third they placed the bismuth cooled system, and fourth they placed the water cooled system. Somewhat later we heard that du Pont felt there was only one chance in a hundred for successfully building a power unit and that they wanted this to be understood in case they should be willing to accept a contract. Thereupon General Groves appointed another committee headed by Lewis of M.I.T., having as a member Mr. Murphree, I believe, of Standard Oil of New Jersey (I am not quite sure whether my memory is correct on this point).[23] All the other members of the committee were employees of du Pont. We were told on November 19, 1942, of the appointment of this committee and were asked to have a report ready by November 23rd. The report was to include three

---

[23]Warren K. Lewis, professor of chemical engineering. His committee included Crawford H. Greenewalt, Tom C. Gary, Roger Williams, and Murphree, who was unable to attend due to illness. See Hewlett and Anderson, *The New World*, p. 110.

pages about the water cooled system and three pages about the bismuth cooling but no more. Shortly afterwards we heard that du Pont was now satisfied that they could build a helium cooled power unit and on that basis was favorably considering the acceptance of a contract from the Government.

After a contract was placed with du Pont Mr. Wigner submitted a scheme for a water cooled power unit in the form of a detailed report. During the second half of January I heard that this system was seriously being considered by du Pont who by that time discovered the difficulties of the helium system, of which the competent men at Chicago had known and written long before. About the middle of February du Pont abandoned the idea of building a helium cooled power unit and had decided to build three water cooled power units of 250,000 kw each along the lines of Mr. Wigner's design. Mr. Wigner urged that in view of the uncertainties of the water cooled system an alternative design should be carried out at least into the process design stage but no such action was taken.

DECISION AS TO SITE

In the assumption that a helium cooled power unit would be built a remote site in the Pacific Northwest was selected because of the availability of large quantities of power that is required by the helium cooled system. What is actually being built at this remote site now is a water cooled production unit which should not require great quantities of electric power. By a lucky coincidence, water of good quality was available at that site so that the false selection of the site did not cause damage in this respect. However, the difficulty of obtaining labor in that area might have ruled out that site had not the false assumption of power requirement induced its acceptance. . . .

DECISION OF UNKNOWN ORIGIN

2. In February 1943 we heard that an air cooled experimental plant of 1,000 kw would be built in the Tennessee Valley.[24] The men whom I contacted in the laboratory did not know how this decision came about. It was everybody's opinion that since du Pont decided to build an industrial plant based on a water cooled system, a water cooled pilot plant of 10,000 or 20,000 kw ought to be built with the utmost speed in place of an air cooled plant of ten times lower capacity. Competent men in the Chicago Laboratory were quite willing to take the responsibility of building this pilot plant or collaborate with du Pont if du Pont cared to build a pilot plant. No pilot plant was built and a $300 million industrial plant is now supposed to go into operation in June 1944 without the benefit of ever having built and operated a pilot plant.

Collaboration Chicago-du Pont

THE QUESTION-ANSWER GAME

I wish to describe now what is commonly called, in the Chicago Laboratory, the

[24] At Oak Ridge, Tennessee.

"question-answer" game. Take for instance the Northwest Pacific construction which will swallow 3/4 million cubic meters of concrete and require the construction of 180 miles of roads and 160 miles of railroad. In this sprawling power plant the first three power units are spaced at 10 miles apart and it will probably be claimed, and in a sense it is probably true, that these large distances were chosen at the recommendations of the Chicago Laboratory. Such recommendations come about through the working of the question-answer game. We may be asked what would be a safe distance to place neighboring power units and if the question is put this way we are inclined to go out of our way and make the most pessimistic conceivable assumptions and then come out with 10 miles as a safe distance to be recommended. If instead of being faced with such a question we had to deal with the problem as a whole and were asked to suggest a reasonable compromise between maximum possible safety and minimum waste of men and materials, which incidentally also coincides with minimum waste of time, the answer would come out very different.

DU PONT NOT ABLE TO UTILIZE OUR STAFF

It is difficult to avoid this question-answer game if two organizations collaborate without integrating their staff into one unified group. When we learned that du Pont decided to drop the "favorite" and build a water cooled system, the whole Laboratory was anxious to cooperate in this work. This, however, did not prove to be feasible. An offer to send Mr. Fermi to Wilmington[25] to help in this work was politely refused. An offer of Mr. Wigner and his group to move to Wilmington to collaborate and help in furthering this design was declined. The last attempt in this direction was made by H. Anderson who went to Wilmington in the assumption that his help was both needed and acceptable. The first assumption was right but the second was not and he is now back in Chicago. . . .[26]

In the circumstances the collaboration of the Chicago Laboratory in designing work is quite unsatisfactory. It is true that the blueprints of the construction are made available to the Chicago Laboratory, but the competent men in this Laboratory say that they are unable to see from these blueprints the whole story and therefore are unable to see whether, from a point of view of the knowledge accumulated in this Laboratory, one ought to expect the construction shown in the blueprints to function properly. . . .

INFLUENCE OF THE DU PONT COMPANY

Since in the present setup the Government cannot be guided by the advice of the scientists, it has to turn for advice to its contractors, who thus play a dual role as both contractors and advisors of the Government. This dual role of the contractors is

---

[25]Wilmington, Delaware, du Pont headquarters.
[26]However, the Princeton physicist John Wheeler did join du Pont to assure liaison between engineers and nuclear physicists.

severely criticized by the scientists who believe that it leads to the suppression of certain promising branches of this work.

So far all those lines of development which utilize the chain reaction and can be expected to survive the war are concentrated in the hands of the du Pont Company. The scientists have come to believe that the du Pont Company picks out certain lines of development for which it happens to be particularly well adapted by its existing equipment or by its available personnel and subsequently they successfully exert an influence in the direction of preventing the scientists from establishing a collaboration with other firms for the development of alternative lines of development (for which the du Pont Company is not suited either because it lacks equipment or because it lacks suitable personnel or because the men in charge of the du Pont Company lack the theoretical background to enable them to appreciate those alternative lines of development).

The first instance of keeping out competitors of du Pont occurred shortly after Thanksgiving 1942. At that time a contract was placed with the du Pont Company for producing heavy water by one of several processes which were proposed by Dr. Urey. The du Pont Company found one of these processes more suitable than the others from the point of view of their previous experience and from the point of view of their equipment.[27] It is certainly the privilege of every company to choose the method which suits them best. Since, however, in the process adopted by du Pont a very large quantity of coal is required for the production of heavy water and since the quantity of heavy water produced by this process is necessarily too small to be of real significance for the production of plutonium, it appeared desirable to develop a more economical process. In the opinion of Urey and his collaborators such an alternative process was available and could be expected to be about three or four times as efficient with respect to coal consumption.[28] In addition they were of the opinion that the investment costs would be considerably lower for this alternative process. Naturally they were anxious to discuss this alternative process with some chemical firm from the production point of view. Their request for permission to do so was refused. The view that this refusal is justified because, in the interest of secrecy, firms other than du Pont must be kept out of the field of heavy water production is not shared by the scientists.

That the scientists are unable to obtain permission to collaborate along a major line in the field of unseparated uranium with a company other than du Pont has become a general conviction by January 1943. This conviction was further borne out by the experience encountered in connection with the development of a heavy water power unit. The history of the efforts of the scientists to develop a design for the heavy water power unit at Chicago is told in the next paragraph. It acquires particular significance because the events connected with it have led to a loss of faith in the

---

[27] Fractional distillation, an expensive and crude but foolproof method.
[28] A sophisticated catalytic exchange method, partly but not completely proved in pilot studies.

management of the Washington end of our organization on the part of a very large number of scientists in the Chicago Laboratory.

DESIGN OF THE HEAVY WATER POWER UNIT

In April 1943 it was decided that a heavy water production unit should be designed and built as speedily as possible. In view of Mr. Wigner's past achievements in connection with the engineering development of the water cooled power unit he was asked to take charge of this work at the Chicago Laboratory. Mr. Wigner felt that if the Laboratory put forward a process design and later a contract for building this power unit were given to an industrial firm, that firm would want to re-design the unit and that a period of something like six months would thereby be lost. For this reason, he asked that the du Pont Company delegate a number of their men to participate in the designing work so that redesigning could be avoided. When it turned out that the du Pont Company had no suitable men available at this time, Mr. Wigner asked to be permitted to approach a company other than the du Pont Company who could collaborate in the designing work and could take over the construction afterwards.

Few scientists in the Laboratory believed that Mr. Wigner's group would succeed in obtaining permission to collaborate with a company other than du Pont. However, in the meantime preparations for this work continued. A large number of men were brought over from Mr. Urey's project to the Chicago Laboratory and a number of men in the Chicago Laboratory were assigned to work under Mr. Wigner on this problem.

In the middle of August 1943 an ad hoc appointed committee under the chairmanship of Lewis came to Chicago and we were informed that this committee would make a final recommendation concerning the organization of the heavy water power unit work.

As a rule ad hoc appointed committees were found by us to be very unsatisfactory. Sometimes these committees are composed of an overwhelming number of men who are employed by a contractor of the Government. In other cases the composition of the committees may not be objectionable on that ground, but rarely is the majority of such a committee acquainted with our work and I know of no other case where the ad hoc appointed committee remained sufficiently long in action to give its members an opportunity to get acquainted with our problems.

In this particular case, however, two of the four members of the committee had previous contact with our work at Chicago. The committee went to great trouble to listen to all points which were brought up by the members of the Chicago Laboratory and also heard Dr. Urey. For the first time the scientists felt that they had been heard by an unbiased committee.

The report of the committee was appreciated by the scientists. It recommended continuation of the designing work on the heavy water power unit and also recommended the immediate establishment of contact with the industrial company that would later be in charge of the construction work. They specifically named a number of industrial companies from which one ought to be chosen for this purpose. The list given by them did not include du Pont. Few men believed, however, that the recommendation of the committee would be carried out and many considered it more likely that no contract for the construction of the heavy water power unit would be placed rather than that the contract would be placed with a firm other than du Pont.

Soon afterwards it was pointed out to us that the heavy water construction was not very important after all, since it was only an insurance against a possible failure of the water cooled graphite pile, which in turn was only an insurance of the possible failure of the hexafluoride diffusion process, which is most likely to give large quantities of the product at an early date.[29]

Next we learned that no contracts would be authorized for the construction of a heavy water power unit. The Chicago Laboratory was permitted to make a design and carry it into the process design stage but no permission was given to contact any of the firms recommended by the committee.

In the meantime, Mr. Wigner's authority in guiding the heavy water designing work was seriously diminished and important responsibilities relating to this designing work were delegated to Mr. Vernon and others (Mr. Vernon is an engineer employed by du Pont who is at present assistant director of the Metallurgical Laboratory). Mr. Vernon put in a request for 80 chemists and engineers and another man put in a request for 50, that is, a total of 130 highly trained men. These requests were forwarded by the Chicago Laboratory without the approval of Mr. Wigner. Shortly afterwards we were informed of a new directive to the effect that all work on the design of a heavy water power unit had to be stopped. This happened in August 1943.

From April to August a large number of men in the Chicago Laboratory were giving their attention to the heavy water power unit and all these men were suddenly thrown out of their track. This sequel of events had an intensely demoralizing effect on those men.

The picture however would not be complete and somewhat unfair to the Washington end of our organization if I did not record one more significant fact. By the time we were informed that the heavy water designing work must be discontinued, Mr. Fermi and Mr. Wigner had reached the conclusion that the defects of the organization of the heavy water designing work in the Chicago Laboratory (which manifested themselves for example in exaggerated requests for highly trained men) were so serious that they would have endangered the success of the designing work. Both Mr. Wigner and Mr. Fermi told me that for that reason they did not wish to express any regret over the total discontinuance of the work.

---

[29]Gaseous diffusion of uranium hexafluoride produced enough uranium-235 for bombs at about the same time plutonium from the Hanford reactors began to appear. The heavy-water program could have been as successful as the Hanford reactors, but was given a lower level of effort and funds.

Part III
Rule by Directives

One of the great weaknesses of the administration of the uranium work consists in suppressing lines of development completely or almost completely which could be carried out with comparatively little effort and expense and which may play an important part in the future. Some of such previously suppressed lines of development have risen to prominence and salvaged some of the projects which otherwise would have failed.

Thus the water cooling suddenly became prominent when it turned out that the favorite, i.e., the helium cooled system, had to be dropped. The number of scientists and engineers working on the water cooled system was then quite suddenly raised from one engineer and a handful of physicists to an enormous staff of physicists, engineers, and chemists. This is an unbalanced way of carrying out development work.

It is in the nature of this work that when a new line of development is started a small sum of money goes a long way toward carrying forward such development and in many cases it would not be possible materially to accelerate the speed of development in the early stages by spending huge sums of money and putting a large number of scientists on the problem. The modest facilities needed for the early stages of new lines should not be withheld.

The rule by directives is based on the assumption that there are some men high above in our organization who possess infinite wisdom based on knowledge and foresight and who can therefore issue directives not only governing 90% of the effort of the uranium projects which would be acceptable, but governing 100% of this effort. If 10% of our effort would be removed from these constantly shifting directives it might easily turn out that this 10% of our efforts produced more results than the 90%. . . .

LACK OF PERMANENT BOARD OF EXPERTS

The scientists who feel the need of developing one or the other side line foreseeing future needs and future difficulties and the possible or probable collapse of the favorite line, have nowhere to turn since there is no permanent board of experts before whom they can put their views. Ad hoc appointed committees would be a poor substitute for such a permanent board even if their recommendations were accepted by the authorities. Clearly only men who give their full time and attention to these complex problems can have a balanced judgment and the confidence of the scientific workers in the various projects. . . .

The Scientists without Representation

There is reason to believe that in the past in a surprisingly large number of cases of impending decisions there was in fact a practically unanimous opinion among the scientists who were competent to judge. This opinion found, however, no expression and has not become a matter of official record. In most other cases when there was a division of opinion the dissenting opinion represented a comparatively small minority.

If there had been a mechanism for putting this collective opinion on record it would have been difficult for the authorities who were responsible for taking far-reaching decisions to make the mistakes which were made because those in authority would have been faced with the choice of following the collective recommendation of the scientists or taking the full responsibility of going against a practically unanimous recommendation.

The importance of this question was recognized by us from the very beginning of this work and therefore as soon as we realized that the work on unseparated uranium might enable us to construct at an early date atomic bombs consisting of element 49 we pressed as hard as we could for the official recognition of a group of competent scientists who could put their opinions on the record.[30] Such a committee was at last appointed under the chairmanship of Dr. Urey in June 1940 and was to act in an advisory capacity to Dr. Briggs. The first meeting of this committee was, however, also its last meeting, since the group was immediately dissolved on the ground that it might lead to criticism if later at a Congressional investigation it should be found that Government funds were expended at the recommendation of this committee which included men who were not American citizens of long standing. . . .

As time went on it became gradually clear that the attitude that first manifested itself with respect to the foreign born scientists extended to practically all competent men and at no time since the dissolution of the first committee of scientists was it possible to have a representation of the competent men who are giving their full time and attention to the uranium work. Such a group of scientists would of course have to cut across the borderlines of the different uranium projects in order to be able to perform a useful function and no such group can therefore function as long as compartmentalization within the uranium work is maintained. It will therefore be necessary to discuss whether or not this compartmentalization of information is justified from the point of view of maintaining secrecy so that we may see whether the gain in secrecy, if any, can compensate for the very great damage which this compartmentalization causes. . . .

Part IV
Psychological Situation of the Scientists

In order to understand this situation one has also to understand the psychological reaction of the scientists which is a very important part of the general picture. The trouble started when, in June 1940, it became evident that it would be impossible to have a body who could speak in the name of the scientists who know most about this

---

[30]See *Doc. 82*. Element "49" was plutonium.

field. At that time nobody knew more about the potential possibilities of the chain reaction in uranium than Fermi, Wigner, and myself. At first when we were excluded from the committee which Dr. Urey tried to organize in June 1940 . . . some of our colleagues argued that it would be possible for us to make our voices heard through the mouths of some of our American colleagues.

The attitude taken toward the foreign born scientists in the early stages of this work had far reaching consequences affecting the attitude towards the American born scientists. Once the general principle that authority and responsibility should be given to those who had the best knowledge and judgment is abandoned by discriminating against the foreign born scientists, it is not possible to uphold this principle with respect to American born scientists either. If authority is not given to the best men in the field there does not seem to be any compelling reason to give it to the second-best man and one may give it to the third- or fourth- or fifth-best men, whichever of them appears to be most agreeable on purely subjective grounds.

Immediately after the dissolution of Urey's committee, Mr. Wigner wrote to Dr. Briggs asking to be relieved from collaborating in the uranium work. He did not take an active part in this work until much later. Since Mr. Wigner was a sort of symbol for those who were convinced of the necessity of collaborating in this field with the U. S. Government, the fact that he was discouraged was an incalculable loss at that time.

When in November 1940 contracts were placed with Columbia University, it was specified that Fermi and I not be given knowledge concerning the development of the centrifuge supported by the Navy. Mr. Fermi was visibly affected by this and he has from that time on shown a very marked attitude of being always ready to be of service rather than considering it his duty to take the initiative. It so happens that early in 1940 I had given some thought to methods of maintaining a counter current of uranium hexafluoride in a centrifuge which I discussed with Urey and after November 1940 Urey was not free to continue these discussions. I felt so discouraged by this that I failed even to work up and write down some earlier nuclear measurements which Zinn and I made in 1939. These measurements gave information on the fission cross section of U^{235}. Had our value been known to Urey and had we been aware that separation of the uranium isotopes would receive adequate support, we certainly would have gone through the simple calculations. . . .

Our British colleagues who visited this country in 1941 fully realized the harm done to the American work through this compartmentalization and freely expressed themselves to this effect. Their attitude was summed up by G. P. Thompson[31] who said something like this: I believe that the Government may succeed in keeping most of the work on uranium secret from the men who do the work—but will it succeed in keeping any of it secret from the Germans?

It would, however, be a mistake to think that in keeping pertinent information

---

[31]George P. Thomson visited the United States in October 1941.

from physicists is an isolated phenomenon limited to the foreign-born. That was the case only when the trouble started; later information was more and more kept from Dr. Urey who is American born, a Nobel prize winner, and from other Americans who have a key position in this work. Among the American Nobel prize winners Dr. Urey has been perhaps pushed around more than any of the others. . . . Some excuse can always be found in every individual case, but the net result is that the scientists are annoyed, feel unhappy and incapable of living up to their responsibility which this unexpected turn in the development of physics has thrown into their lap. As a consequence of this, the morale has suffered to the point where it almost amounts to a loss of faith. The scientists shrug their shoulders and go through the motions of performing their duty. They no longer consider the overall success of this work as their responsibility. In the Chicago project the morale of the scientists could almost be plotted in a graph by counting the number of lights burning after dinner in the offices of Eckhart Hall. At present the lights are out.

Secrecy

I wish now to discuss what gain in secrecy is achieved by withholding information from those scientists who have shown initiative in the past and who would show initiative at present if they were not frustrated in their work. There is a unanimous opinion among the scientists that there is no appreciable gain in secrecy by withholding information from the key men. All those men have a long record which is known to many of their colleagues. Many of them have know each other for over 20 years. It is inconceivable that any of these scientists should disclose technical information to the enemy. As a matter of fact, there is not a single case in the history of the world on record where a scientist has betrayed the trust of the Government for which he worked in wartime.

That secrecy is important was recognized by many of us in March 1939 and there is a long record of documents showing that we did our best to get the support of the Government for organizing secrecy and that it was the representatives of the Government who failed to realize the importance of secrecy at that time, just as later on they failed to realize the harm due to secrecy in the wrong place.

In the circumstances the compartmentalization imposed on the scientists is considered by them as unjustified and the argument that it is necessary for reasons of security is rejected by them.

At present the responsibility for secrecy rests with the Army and the methods for safeguarding secrecy are essentially the routine methods which usually are applied by the Army to developments very different from the character of the uranium projects. Leaks at present are numerous and the nature of some of those leaks is well known to the scientists. Few of us doubt, for instance, that the Germans know the location of our sites, which is very regrettable.[32] No information of a technical nature is, however,

---

[32]In fact the Germans were ignorant of nearly all aspects of the American bomb project; but the Russians knew a great deal.

leaking out through the scientists who have key positions in this work, at least none of us believes that leaks of this nature occur.

Objective Losses due to Compartmentalization

The most important loss due to compartmentalization of information in the uranium work is probably represented by the methods which remain undiscovered in America in spite of the large number of potential inventors engaged in the uranium work because of the frustration of the scientists due to the compartmentalization. This loss cannot be objectively evaluated at present. In the following I will list, however, a number of instances of which I have intimate personal knowledge of the facts and am satisfied that compartmentalization of information led to a considerable delay of the work:

1. The failure to realize the possibility of small and efficient atomic bombs built of 25 mentioned before.[33]

2. Our failure to introduce the magnesium reduction of uranium tetrafluoride in 1941. This was primarily due to the fact that the knowledge that uranium tetrafluoride was available in Urey's project was withheld from me and so I was not able to take any action along this line until the middle of 1942. This caused both a great loss in time and a substantial loss in money. . . .

5. Another instance where secrecy in the wrong place has caused considerable damage is connected with the development of the centrifuge by Westinghouse. The men at Westinghouse who knew the portent of that development were not free to communicate their knowledge to certain other key men in the Westinghouse organization upon whose good will the facilities made available for this development by Westinghouse depended. This slowed down the development of the centrifuge by Westinghouse. This and also that Murphree's men failed adequately to realize the importance of the counter current method sponsored by Urey, accounts probably for the fact that the centrifuge development fell behind so much that it practically was thrown overboard in favor of other developments. The scientists do not know enough of these happenings to be able to express any opinion with assurance, but they are far from being satisfied that the centrifuge is not a potent method for separating isotopes. It would not be too surprising to me if atomic bombs manufactured by the Germans by the centrifuge method were the first to go into action.

[33] i.e., built of uranium-235.

Chapter VI
Some of us began to think about the wisdom of testing bombs and using bombs.

After it became clear that the Hanford plants were successfully operating, the Chicago project relaxed. It then became possible for the physicists to take a more detached view, and some of us began to think about the wisdom of testing bombs and using bombs. [*10*] In the spring of '45 it was clear that the war against Germany would soon end, and so I began to ask myself, "What is the purpose of continuing the development of the bomb, and how would the bomb be used if the war with Japan has not ended by the time we have the first bombs?" (*Doc. 99*)

Initially we were strongly motivated to produce the bomb because we feared that the Germans would get ahead of us, and the only way to prevent them from dropping bombs on us was to have bombs in readiness ourselves. (*Doc. 100*) But now, with the war won, it was not clear what we were working for.

I had many discussions with many people about this point in the Metallurgical Laboratory of the University of Chicago (which was the code name for the uranium project that produced the chain reaction). There was no indication that these problems were seriously discussed at a high government level. I had repeated conversations with Compton about the future of the project, and he too was concerned about its future, but he had no word of what intentions there were, if there were any intentions at all.

There was no point in discussion these things with General Groves[1] or Dr. Conant[2] or Dr. Bush, and because of secrecy there was no intermediate level in the government to which we could have gone for a careful consideration of these issues. The only man with whom we were sure we would be entitled to communicate was the President. In these circumstances I wrote a memorandum addressed to the President and was looking around for some ways and means to communicate the memorandum to him. Since I didn't suppose that he would know who I was, I needed a letter of introduction.

I went to see Einstein and I asked him to write me such a letter of introduction, even though I could tell him only that there was trouble ahead but I couldn't tell him what the nature of the trouble was. Einstein wrote a letter and I decided to transmit the memorandum and the letter to the President through Mrs. Roosevelt, who once before had channelled communications from the project to the President.[3] I have forgotten now precisely what I wrote Mrs. Roosevelt; I suppose that I sent her a copy of Einstein's letter, but not the memorandum. The memorandum I couldn't send her, because the memorandum would have been considered secret. (*Docs. 101–102*)

Mrs. Roosevelt gave me an appointment [for May 8, 1945]. When I had this ap-

[1] Major General Leslie R. Groves, director of the Manhattan Project.
[2] James B. Conant, president of Harvard University and chairman of the National Defense Research Committee, which oversaw fission work under Bush's overall direction.
[3] See n. 13, p. 162.

pointment I called on Dr. Compton, who was in charge of the project, and told him that I intended to get a memorandum to the President, and I asked him to read the memorandum. I was fully prepared to be scolded by Compton and to be told that I should go through channels rather than go to the President directly. To my astonishment, this is not what happened. Compton read the memorandum very carefully, and then said,[4] "I hope that you will get the President to read this." Elated by finding no resistance where I expected resistance, I went back to my office. I hadn't been in my office for five minutes when there was a knock on the door and Compton's assistant came in, telling me that he had just heard over the radio that President Roosevelt had died [April 12, 1945].

There I was now with my memorandum, and no way to get it anywhere. At this point I knew that I was in need of advice. I went to see the associate director of the project, Dr. [Walter] Bartky, and told him of my plight. He suggested that we go and see Dr. [Robert M.] Hutchins, president of the University [of Chicago]. This was the first time that I met Hutchins. I told him briefly what the situation was, and this was the first time that he knew that we were close to having an atomic bomb, even though the Metallurgical Project had been on his campus for several years. Hutchins grasped the situation in an instant. He used to be an isolationist before the war, but he was a very peculiar isolationist, because where most isolationists held that the Americans should keep out of war because those foreigners did not deserve to have American blood shed for them, Hutchins' position was that the Americans should keep out of war because they would only mess it up. After he heard my story he asked me what this would all mean in the end, and I said that in the end this would mean that the world would have to live under one government. Hutchins said, "Yes, I believe you are right." I thought this was pretty good for an isolationist. As a matter of fact, a few days after the bomb was dropped on Hiroshima, Hutchins went on the radio; he gave a speech about the necessity of world government.[5]

In spite of the good understanding which I had with Hutchins, he was not able to help with the task immediately at hand. "I do not know Mr. Truman," Hutchins said. I knew any number of people who could have reached Roosevelt, but I knew nobody offhand who could have reached Truman; Truman just did not move in the same circles. So for a number of days I was at a complete loss for what to do. Then I had an idea. Our project was very large by then, and there ought to be somebody from Kansas City. So I looked around, and sure enough there was someone from Kansas City; and three days later we had an appointment at the White House. (*Doc. 103*)

I asked the associate director of the project, Dr. Bartky, to come with me to Washington. Armed with Einstein's letter and my memorandum, we went to the White House and were received by Matt Connelly, Truman's appointments secretary. I handed him Einstein's letter and the memorandum to read. He read the memorandum

---

[4]On the tape Szilard says, "and then I said . . ." We assume this was a slip of the tongue.
[5]See University of Chicago *Round Table*, no. 386 (August 12, 1945).

carefully from beginning to end, and then he said, "I see now, this is a serious matter. At first I was a little suspicious, because this appointment came through Kansas City." Then he said, "The President thought that your concern would be about this matter, and he has asked me to make an appointment for you with James Byrnes, if you are willing to go down to see him in Spartanburg, South Carolina." We said that we'd be happy to go anywhere that the President directed us to go, and he picked up the phone and made an appointment with Byrnes for us. I asked whether I might bring Dr. H. C. Urey along, and Connelly said I could bring along anyone whom I wanted. So I phoned Chicago and asked Urey to join us in Washington, and together we went down the next day to Spartanburg, taking an overnight train from Washington.

We were concerned about two things. We were concerned first about the role which the bomb would play in the world after the war, and how America's position would be affected if the bomb were actually used in the war. We were also concerned about the future of atomic energy, and about the lack of planning of how this research might be continued after the war. It was clear that the projects set up during the war would not be continued but would have to be reorganized. But the valuable thing was not the big projects; the valuable things were the numerous teams, which somehow crystallized during the war, of men who had different abilities and who liked to work together with each other. We thought that these teams ought to be preserved even though the projects might be dissolved.

We did not quite understand why we were sent by the President to see James Byrnes. Byrnes had occupied a high position in the government, but was now out of the government and was living as a private citizen in Spartanburg. Clearly the President must have had in mind to appoint him to a government position, but to what position? Was he to [be] appointed to be the man in charge of the uranium work after the war, or what? We did not know.

Finally we arrived in Spartanburg and I gave Byrnes Einstein's letter to read and the memorandum which I had written. Byrnes read the memorandum, and then we started to discuss the problem. When I spoke of my concern that Russia might become an atomic power, and might become an atomic power soon, if we demonstrated the power of the bomb and if we used it against Japan, his reply was, "General Groves tells me there is no uranium in Russia."

I told Byrnes that there was certainly a limited amount of rich uranium ore in Czechoslovakia to which Russia had access; but apart from this, it was very unlikely that in the vast territory of Russia there should be no low-grade uranium ores. High-grade uranium ore is of course another matter: high-grade deposits are rare, and it was not sure that some new high-grade deposits could be found. In the past only the high-grade deposits were of interest, because the main purpose of mining uranium ores was to produce radium, and the price of radium was such that working low-grade uranium ores would not have been profitable. But when you are dealing with atomic energy you are not limited to high-grade ores; you can use low-grade ores. I doubted very much that anyone in America would be able to say, in a responsible way, that there were no major low-grade uranium deposits in Russia.

I thought that it would be a mistake to disclose the existence of the bomb to the world before the government had made up its mind about how to handle the situation after the war. Using the bomb certainly would disclose that the bomb existed. As a matter of fact, even testing the bomb would disclose that the bomb existed. Once the bomb had been tested and shown to go off, it would not be possible to keep it secret.

Byrnes agreed that if we refrained from testing the bomb people would conclude that the development of the bomb had not succeeded. However, he said we had spent two billion dollars on developing the bomb, and Congress would want to know what we had got for the money spent. He said, "How would you get Congress to appropriate money for atomic energy research if you do not show results for the money which has been spent already?"

I saw his point at that time, and in retrospect I seen even more clearly that it would not have served any useful purpose to keep the bomb secret, waiting for the government to understand the problem and to formulate a policy. For the government will not formulate a policy unless it is under pressure to do so, and if the bomb had been kept secret there would have been no pressure for the government to do anything in this direction.

Byrnes thought that the war would be over in about six months (this proved to be a fairly accurate estimate). He was concerned about Russia's postwar behavior. Russian troops had moved into Hungary and Rumania, and Byrnes thought it would be very difficult to persuade Russia to withdraw her troops from these countries, that Russia might be more manageable if impressed by American military might, and that a demonstration of the bomb might impress Russia. I shared Byrnes' concern about Russia's throwing around her weight in the postwar period, but I was completely flabbergasted by the assumption that rattling the bomb might make Russia more manageable.

I began to doubt that there was any way for me to communicate with Byrnes in this matter, and my doubt became certainty when he turned to me and said, "Well, you come from Hungary—you would not want Russia to stay in Hungary indefinitely." I certainly didn't want Russia to stay in Hungary indefinitely, but what Byrnes said offended my sense of proportion. I was concerned at this point that by demonstrating the bomb and using it in the war against Japan, we might start an atomic arms race between America and Russia which might end with the destruction of both countries. I was not disposed at this point to worry about what would happen to Hungary.

After all was said that could be said on this topic, the conversation turned to the future of the uranium project. To our astonishment, Byrnes showed complete indifference. This is easy to understand in retrospect, because contrary to what we had suspected, he was not slated to be director of the uranium project; he was slated to be Secretary of State.

I was rarely as depressed as when we left Byrnes' house and walked toward the station. I thought to myself how much better off the world might be had I been born in America and become influential in American politics, and had Byrnes been born in Hungary and studied physics. In all probability there would then have been no atomic bomb and no danger of an arms race between America and Russia.

On my way from Spartanburg to Chicago I stopped in Washington to see Oppenheimer, who had arrived there to attend a meeting of the Interim Committee.[6] I told Oppenheimer that I thought it would be a very serious mistake to use the bomb against the cities of Japan. Oppenheimer didn't share my view. He surprised me by starting the conversation by saying, "The atomic bomb is shit." "What do you mean by that?" I asked him. He said, "Well, this is a weapon which has no military significance. It will make a big bang—a very big bang—but it is not a weapon which is useful in war." He thought that it would be important, however, to inform the Russians that we had an atomic bomb and that we intended to use it against the cities of Japan, rather than taking them by surprise. This seemed reasonable to me, and I knew that Stimson[7] also shared this view. However, while this was necessary it was certainly not sufficient. "Well," Oppenheimer said, "don't you think that if we tell the Russians what we intend to do and then use the bomb in Japan, the Russians will understand it?" And I remember that I said, "They'll understand it only too well."[8] [2]

This [Interim] Committee met for the first time after I had seen Byrnes. Stimson was chairman of the committee. Byrnes attended only a few of the meetings. I was very unhappy about the composition of the committee, because many of the people had a vested interest that the bomb be used. You see, we had spent two billion dollars. Bush and Conant felt a responsibility for having spent two billion dollars and they would, I think, have very much regretted not to have something to show for it. The Army was determined to drop the bomb. Again, they wanted to show that they did it and did it successfully. Stimson I have the highest respect for. I think Stimson was the most thoughtful of the Interim Committee. But when finally you read in his own words why he decided to drop the bomb, you see that he had no case. He wrote, I think in *Atlantic Monthly*,[9] the following: that we could not demonstrate the bomb, because we had only two bombs, and these could have been duds. If we had demonstrated the bombs and they had proved to be duds, we would have lost face. Now this was a completely invalid argument. It was true we had only two bombs. [But] it would not have been necessary to wait for very long before we would have had enough bombs to eliminate the risk that they were all duds.[10] [*11*]

---

[6]The Interim Committee was organized in early May of 1945 to consider uses of the bomb and possible international control. J. Robert Oppenheimer, director of the Los Alamos laboratory, was on the scientific advisory panel to the Interim Committee. Hewlett and Anderson, *The New World*, 344–346.
[7]Secretary of War Henry L. Stimson.
[8]In the original tapes this paragraph comes later, out of chronological order; see below, n. 12.
[9]Henry L. Stimson, "The Decision to Use the Atomic Bomb," *Harper's, 194*: 97–107 (February, 1947).
Stimson's main argument was that it was the threat of further bombs (not yet assembled) that caused the Japanese surrender.
[10]On another occasion Szilard said, "It would have taken a very short time until we would have had ten bombs." [*13*]

Four physicists, namely Fermi, Lawrence, Compton and Oppenheimer, were chosen to represent the scientists before this Committee. The composition of the Committee and the selection of the physicists disturbed us, for while the physicists were all good men, they were men who could be expected to play ball on this occasion. Oppenheimer, we thought, would not oppose the using of the bomb which he had tried so hard to make. Fermi would state his opinion but would not insist that it should be heard, and would not state it a second time. Compton might be against the use of the bomb but he would not want to incur the displeasure of the powers by stressing this point of view. And of Lawrence's position we knew too little to be comforted. [10]

When I returned to Chicago I found the project in an uproar. The Army had violently objected to our visit to the White House and to Byrnes. Dr. Bartky was summoned to see General Groves; General Groves told him that I had committed a grave breach of security by handing a secret document to Byrnes, who did not know how to handle secret documents. To calm the uproar, Dr. Compton, the leader of the project, decided to regularize the discussions by appointing a committee, under the chairmanship of James Franck,[11] to examine the issue of whether or not the bomb should be used, and if so, how.[12] [2] The report of this committee was rushed to Stimson and advised against the outright military use of atomic bombs in the war against Japan. It took a stand in favor of demonstrating the power of the atomic bomb in a manner which would avoid mass slaughter but yet convince the Japanese of the destructive power of the bomb.[13] [12]

I think it is clear that you can't demonstrate a bomb over an uninhabited island. You have to demolish a city. So a demonstration would have meant approaching Japan through a diplomatic channel, proposing a demonstration, say, over Hiroshima with the inhabitants removed from Hiroshima. In retrospect, I think that the discussions of the demonstration overemphasized the need for a demonstration. What we did not discuss enough was that Japan was defeated; the war could be ended by political means and need not be ended by military means. [13]

The time approached when the bomb would be tested. The date was never communicated to us in Chicago, nor did we ever receive any official indication of what was afoot. However, I concluded that the bomb was about to be tested when I was told that we were no longer permitted to call Los Alamos over the telephone. This could have meant only one thing: Los Alamos must be getting ready to test the bomb, and the Army tried by this ingenious method to keep the news from the Chicago project.

---

[11]Emigré physicist at the Chicago Laboratory.

[12]There follows a paragraph that has been put elsewhere to preserve chronological order; see above n. 8.

[13]Szilard was a member of the committee. Eugene Rabinowitch has stated that the text "was prepared essentially by me with the important contribution of Leo Szilard. . . [who] was responsible for the whole emphasis on the problem of the use of the bomb . . . the attempt to prevent the use of the bomb on Japan." Herbert Feis, *The Atomic Bomb and the End of World War II* (Princeton: Princeton University Press, 1966), p. 51. The text of the final version is printed in Morton Grodzins and Eugene Rabinowitch, eds., *The Atomic Age. Scientists in National and World Affairs* (New York: Simon & Schuster, 1963), pp. 19–27.

I knew by this time that it would not be possible to dissuade the government from using the bomb against the cities of Japan. The cards in the Interim Committee were stacked against such an approach to the problem. Therefore all that remained to be done was for the scientists to go unmistakably on record that they were opposed to such action. While the Franck Report argued the case on the ground of expediency, I thought that the time had come for the scientists to go on record against the use of the bomb against the cities of Japan, on moral grounds. Therefore I drafted a petition which I circulated in the project.[14]

This was again violently opposed by the Army. They accused me of having violated secrecy by disclosing in the petition that such a thing as a bomb existed. What the Army thought that we thought we were doing all this time, I cannot say. However, we did not yield to the Army's demand. The right to petition is anchored in the Constitution, and when you are a naturalized citizen you are supposed to learn the Constitution prior to obtaining your citizenship.

The first version of the petition which was circulated drew about fifty-three signatures in the Chicago project. What is significant is that these fifty-three people included all the leading physicists in the project and many of the leading biologists. The signatures of the chemists were conspicuously absent. This was so striking that I went over to the chemistry department to discover what the trouble was. What I discovered was rather disturbing: the chemists argued that what we must determine was solely whether more lives would be saved by using the bomb or by continuing the war without using the bomb. This was a utilitarian argument with which I was very familiar through my previous experiences in Germany. That some other issue might be involved in dropping a bomb on an inhabited city and killing men, women, and children did not occur to any of the chemists with whom I spoke.

Some of the members of the project said that they would sign the petition if it were worded somewhat more mildly, and I therefore drafted a second version of the petition which drew a somewhat larger number of signatures—but not a significantly larger number. The second petition was dated one day before the bomb was actually tested at Alamogordo, New Mexico.[15] (*Doc. 104–109*)

After the petition had been circulated we were faced with the decision of through what channel to communicate it to the White House. Several people, and above all James Franck, took the position that they would sign the petition because they agreed with it, but they could do this only if the petition would be forwarded to the President through the regular channels rather than directly, outside of these channels. I did not like this idea because I was just not sure whether the regular channels would forward the petition or whether they would sabotage it by filing it until the war was

---

[14]Szilard wrote earlier, "A petition to the President was thus drafted in the first days of July and sent to every group leader in the 'Metallurgical Laboratory,' with the request to circulate it within his group." [*12*]

[15]The Alamogordo test was July 16, 1945. We have not found any version dated July 15. There is one dated the 16th, the day of the test, which is almost identical to the July 17 version but without any signatures. All of the copies with signatures are dated either the 3rd or the 17th.

over. However, to my regret I finally yielded and handed the petition to Compton, who transmitted it to Colonel Nichols, who promised that he would transmit it to General Groves for immediate transmittal to Potsdam. (*Docs. 110–111*) I have no evidence that the petition ever reached the President.[16] [2]

I knew that the bomb would be dropped, that we had lost the fight. And when it was actually dropped my overall feeling was a feeling of relief. A component of this relief is that we were completely bottled up in our discussions—it was not possible to get real issues before the public because of secrecy. Suddenly the secrecy was dropped and it was possible to tell people what this was about and what we were facing in this century. [11]

After the bomb had been dropped on Hiroshima, I called the responsible officer of the Manhattan District in Chicago and told him that I was going to declassify the petition and asked him whether there was any objection. There could not have been any objection, and there wasn't, and so I declassified the petition. A short time thereafter I sent a telegram to Matt Connelly, the President's secretary, to advise him that it was my intention to make the contents of the petition public, and that I wanted to advise him of this as a matter of courtesy.[17] When the telegram was not acknowledged I phoned the White House, upon which I received a telegram saying that the matter had been presented to the President for his decision, and that I would be advised accordingly. Shortly thereafter I received a call from the Manhattan District, saying that General Groves wanted the petition to be reclassified "Secret." I said that I would not do this on the basis of a telephone conversation, but I would want to have a letter explaining for what reason the petition, which contained nothing secret, should be reclassified. Soon after I received a three-page letter, stamped "Secret," in which I was advised that while the officer writing the letter could not possibly know what was in General Groves' mind when he asked that the petition be reclassified "Secret," he assumed that the reason for this request was that people reading the petition might conclude there must have been some dissension in the project prior to the termination of the war, which might have slowed down the work of the project which was conducted under the Army. [2] (*Docs. 112–116*)

---

[16]The petition never reached President Truman. Groves' assistant Colonel K. D. Nichols delivered the petition on July 25 to Groves, who kept it until August 1, when it was delivered to Stimson's office by messenger. But Truman was then at the Potsdam conference, about to embark for home. On August 6, the day of Hiroshima, he was still on the Atlantic. Almost a year later, May 24, 1946, Army Lieutenant R. Gordon Arneson, secretary of the Interim Committee, wrote in a memorandum that ". . . since the question of the bomb's use 'had already been fully considered and settled by the proper authorities,' . . . it was decided that 'no useful purpose would be served by transmitting either the petition or any of the attached documents to the White House, particularly since the President was not then in the country.' " Fletcher Knebel and Charles W. Bailey, "The Fight over the A-Bomb; Secret Revealed after 18 years," *Look*, *27*: 22–23 (August 13, 1963).

[17]We have not found this telegram to Connelly, but have found a confirming letter, Szilard to Connelly, August 17, 1945.

Documents from August 1944 through August 1945

We are not certain that the following typed memorandum was written by Szilard; conceivably it was drafted by some colleague. But it agrees with what we know of Szilard's thoughts at this time and bears revisions and marginal notes in his hand. It is a draft of a memorandum intended to be sent, as an accompanying summary states, "to those physicists and chemists upon whose discoveries, inventions, initiative, and correct judgment most of the successes of our present work are based. It puts down those considerations and opinions which I believe might be shared by the majority of this group." It was apparently never completed and circulated.

Document 99

August 10, 1944

There has been lately much talk about the organization of our work in peacetime, about such questions as the publication of the results obtained during the war and peacetime application of our discoveries. Some physicists and chemists who may be considered as key men in this work have been offered postwar positions by various firms that want to get into this new field after the war and one of these men has actually accepted such a position. The situation is such that it seems to me very desirable that we should try to agree on a common line of basic policy which could be followed individually and, if necessary, collectively.

The purpose of this letter is to find out whether there is sufficient agreement among us on certain basic principles to formulate such a common policy. I shall in the following therefore record those of my beliefs which I hope I share with the majority of our group and I would very much appreciate it if you would comment on these points.

1. The "modern" development with which we are all familiar will make it impossible to ensure peace on the basis of ideas which date back to 1918 and are essentially characterized by the slogans of cooperation between sovereign states bound together by covenants possibly reinforced by alliances and automatic[1] collective action against nations. Those ideas might have been effective after the *last* war if they had been acted upon at that time but if after *this* war we have an armed peace in which a number of nations are in possession of "modern" weapons[2] and if one of these nations starts a war, that war may be won in 24 hours and there will be no time to apply sanctions. If after this war we get an armed peace based on balance of power of several states in possession of the "modern" weapons then we physicists will have to regard that era as a pre-war rather than a postwar period. Judging from newspaper reports it is such an armed peace at best stabilized by alliances and league of nation covenants towards which we are drifting.

---

[1] The word "automatic" is written into the typescript in an unknown hand.
[2] i.e., nuclear weapons.

It is not entirely out of the question that the physicists and chemists who are engaged in this work will have to raise their voices against the postwar world organization towards which we are drifting at present. Of course it may be that one of these days newspaper headlines will announce the use of "modern" weapons by the Germans on Cherbourg,[3] London, or New York and if that should happen there may be an abrupt change in the present trend and it may not take any explaining by the physicists to make statesmen understand the situation which the world faces.

On the other hand it may also happen that the postwar organization of the world will begin to take shape in Europe before the full meaning of the development of modern weapons has become manifest. In that case the physicists and chemists might perform a useful function by explaining in the right place and at the proper time the grave consequences to which the modern discoveries will lead if we drift into an armed peace. We may take action individually or collectively. As far as collective action is concerned it seems that we can not very well go farther than to state as clearly as we can the necessary conditions for safeguarding peace in face of this modern development.

One such condition would appear to be that all pertinent deposits[4] be controlled by one single closely knit group. If this is to be effective it will require in a sense policing of the whole world so that any danger to security arising out of the violation of the control measures could be met by police action against the administration of the territory which is involved rather than by waging war against the people who happen to inhabit that territory. As to the question of what kind of group could in fact exercise such control it is hardly possible to give an answer.

Theoretically speaking this group could be the United States alone or the United States and Great Britain, if tied together by an indissoluble alliance approaching some sort of a lasting political union. It could be the United States, Great Britain and Russia if it were possible to create an atmosphere in which a lasting union of these three countries could be established. It could be some sort of a League of Nations if it were possible to convince the nations to give up sovereignty to the extent needed to make their union effective.

Any measures of this type which we could name as necessary conditions for peace are clearly not sufficient and we all know that it takes more than such measures to create a peace in which we may have confidence. Unfortunately the public discussions of this topic have not yet reached a sufficiently high level and I believe therefore that it would be premature for us to try to reach a consensus of opinion on this wider subject.

I believe that most of us would agree that the hopes that we shall in fact get a satisfactory organization of the world after this war are slim and that even if the United States Government had a clear determination in this respect and even if the people of the United States were backing the Government 100% in this attempt, it

---

[3]In August, 1944, this was expected to be a major supply port for the invasion of France.
[4]Of uranium and thorium ores.

might still not be possible to obtain such a world organization in view of the attitude which the other countries might take. In spite of this hope being slim, it would I believe be necessary at least to make an attempt in this direction.

The first question which arises then is this: Do we feel responsible for putting forward our views at the proper place and proper time to the extent of making clear the necessary conditions for maintaining peace in the face of the present development? Obviously some allowances will have to be made for future developments and there is hardly anybody in a better position to appreciate the importance of this aspect than our group. Are we willing if necessary to go as far as seeing personally men in key positions in Washington and also seeing Senators and Congressmen if need be, in support of such policies of the administration which might be opposed by those who were not given enough time fully to understand the implications of the developments of which our group of physicists and chemists has been fully aware for the last three years. It appears to me conceivable that a lobby of this unprecedented type might not remain without effect because politicians are usually shrewd and would have no difficulty in seeing that we are sincere and have no political axe to grind. . . .[5]

2. . . . I fear that the chances for a stable peace after this war are slim seeing that it takes time for people to grasp the full meaning of this new development.

For this reason it seems to me that we ought to proceed on the assumption that only an overwhelming superiority of this country, perhaps in collaboration with Great Britain, in the quantity and quality of modern weapons can safeguard peace.

Thus the maintenance of the peace will probably be by far the most important peace application of the "modern development" in the next few years and it is difficult to think of any other peace application which might achieve comparable importance.

It seems to follow that in such a situation we cannot think of returning to the normal method of publications which would make basic information available to the potential enemy. Clearly we have to steer clear of two dangers: One, to have no secrecy at all and give away the information to the enemy, and the other, to hinder the work of our own physicists and chemists by preventing them from freely discussing their results with each other. If we do the former, countries which start from scratch can catch up with us and surpass us within two or three years.[6] If we do the latter, it may take a country starting from scratch somewhat longer to accumulate the basic information needed for this work, but if they trust their own men, while we distrust ours, *they* will be in a much better position with respect to unrecognized problems and therefore in the long run (say 5 to 10 years) far better off than we are. This has been demonstrated: England.[7] It follows that we have a difficult task (1) to persuade our politically less interested physicists to cooperate in an airtight scheme of withholding information from everybody who does not take responsibility to keep this information from leaking out, and (2) to make it clear to "the authorities" that *a large group* of scientists

---

[5]We omit two repetitious paragraphs.
[6]Marginal note by Szilard: "Our advantages: head start. Larger group of scientists."
[7]Szilard felt that the English-sponsored group at Chalk River, Canada, had independently come up with some techniques superior to American ones.

freely communicating information to each other is *a minor* danger to security but that the existence of such a group is *essential* for a well-balanced development . . . .[8]

As to what is secret from the potential enemy it seems to me that we should take the stand that the most important secrets are the basic information, the points of view which we consider ought to be guiding, and the emphasis which we wish to put on one or the other line of development. Once this information is public any country which wishes to start from scratch can with the help of a capable group of engineers solve the remaining constructional problems in an astonishingly short time.

3. The question arises as to whether this development is best carried out under Government auspices or under other institutions where the profit object plays no role or whether industrial research laboratories are more suitable for a speedy development. In examining this question we shall restrict ourselves to fundamental research and that very important region between fundamental research and actual constructional work which consists of a balanced mixture of technological development and designing work and which carries the development from the early discoveries into the process design stage. I personally believe that this part of the development work is best kept out of the hands of commercial companies provided that certain essential requirements can be met if the guidance of the work is in public rather than in private hands. . . .

In the remainder of the memorandum (five of the thirteen pages) the author lists the dangers of public organization as "overcentralization," "decentralization" (by compartmentalization or geographical fragmentation), and "stuffed-shirt direction" (overuse of committees of prestigious but uninformed people, anxiety about making a good record, etc.). Remedies proposed include giving proven achievers on the staff twenty percent of all facilities to follow lines of their own choice, and forming a "panel" or "team" of men closely connected with the scientific work who would gather evidence, discuss future plans, and communicate the results of their deliberations to all administrative authorities.

The following letter shows Szilard's continuing fear that the Germans would soon have a nuclear bomb, and a beginning of serious concern over the effects of such bombs.

Document 100

1155 East 57th Street
Chicago, Illinois
August 18, 1944

Lord Cherwell[9]
10 Downing Street
London, England

Dear Lord Cherwell:

It seems to me that we have to reckon with the possibility that the Germans might now

---
[8]We omit one paragraph elaborating on this.
[9]F. A. Lindemann, then Churchill's scientific adviser.

Chapter VI: Documents from August 1944 through August 1945   193

start using atomic bombs before long and I am writing to you because I hope that you might decide to give this possibility your continued personal attention from now on. Perhaps you are doing this already.

Private communications originating from Switzerland[10] which reached me two years ago indicated that the Germans knew by the middle of 1942 how to make a chain reaction go and that Heisenberg, who was in charge of that work, had some conspicuous successes along this or the other parallel line of work during that year and consequently was put in full charge of *all* the work (made director of the Kaiser-Wilhelm Institute for Physics) late in 1942.[11] This was about the time of Stalingrad after which the Germans must have realized that they may have to win this war by other than ordinary methods. Unless I completely misjudge the psychology of the Germans, they must have gone *full scale* into this work soon after Stalingrad at the latest.

I could hardly hope to convince anybody by this type of reasoning who does not know Germany and the Germans from personal experience. But you yourself lived quite a long time in Germany before 1914 and it might therefore carry weight with you.

We know that there are a *number of methods* which can be used for obtaining materials suitable for atomic bombs. If we rule out the three most expensive methods on the ground that they would have put too great a burden on the already strained German war economy there still remain two methods available to the Germans which are considerably less expensive, and we know that the Germans have been considering them early in 1942. If they have chosen either or both of them and started in earnest after Stalingrad at a scale of a twenty million pound enterprise they ought to be making by now one or two bombs per month.[12]

Of course we cannot entirely exclude the possibility that the Germans have hit on some better methods than those of which we are aware and if so, they may produce a large number of bombs and win the war. However, it seems to me better not to contemplate this possibility but rather to steer a middle course and to assume that they do not know much more than we do and that there is not anything that we know that they do not know also.

If we are willing to proceed on this assumption, we may then regard as the chief danger not so much that the Germans may use the bombs as a military weapon but rather that they may use it as a political weapon. Unless we are fully prepared to meet this danger they may detonate their first bomb high above Whitehall at the time when the House of Commons meets and kill a large number of persons who are important

---

[10]Possibly from E. Dessauer, relayed in a letter from G. Dessauer to Szilard, July 6, 1942. The letter said the German physicist W. Gentner knew that a nonexplosive chain reaction was possible.

[11]The physicist Werner Heisenberg held a strong position but was not the sole leader of the fragmented German program.

[12]Perhaps the three expensive methods were separation of uranium-235 by electromagnetic, gaseous diffusion, or centrifuge plants, and the cheap ones were production of plutonium in graphite or heavy water reactors. Or Szilard may have (wrongly) considered centrifuges to be cheaper than graphite. The Germans followed only the centrifuge and heavy water lines, but devoted few resources to the project.

for the functioning of the machinery of the government. They may detonate a few more bombs over other cities and try to make the world believe that they have an unlimited supply. Unless we are reliably informed of the scope of their industrial effort they may successfully bluff us and reach their political objectives.

I believe that if you become fully acquainted with the scope of action of atomic bombs you will want to acquaint yourself with the material relating to such activity in Germany that must have gradually been gathered by the British Military Intelligence. I should not be surprised however if the British Intelligence had failed so far to find any evidence for the manufacture of the pertinent materials in Germany. Clearly there are a number of possible pitfalls in the way of finding such evidence particularly if the *routine methods* of gathering intelligence are applied to this task. If the agents are not *fully aware of all potential possibilities* they may not find anything unless the questions put to them by the physicists have been selected with great care and circumspection. There may be too much reluctance to tell the agent *enough* to enable him to do his work and the questions put to him may not be pertinent ones. The physicists who collaborate may not be sufficiently convinced that *something* must be going on and therefore may not be able to put together the pieces of the puzzle and obtain a coherent picture from the odd bits of information which may come in.

You might, if you look through the whole record of the information which has been gathered by the British Intelligence, perhaps come to the conclusion that the methods used are capable of improvement or that the *scale* of the investigation ought to be *greatly* increased.

I wonder whether it would be possible for you to organize a small group of physicists either in your office or perhaps in the office of the Prime Minister and to use some members of this group as liaisons with the British Intelligence. If these men are then authorized to report back to you *directly*, you would be able to follow closely the material which is being gathered and be able to judge the proper time when sufficient evidence has been obtained for bringing the matter up before the War Cabinet for such *urgent* action as may then be needed.

I am rather convinced that a properly organized effort of the British Intelligence will lead to the discovery of industrial installations for the manufacture of the relevent materials in Germany but when the location of the German factories is discovered, it may be found that we would have to pay an exceedingly high price for their destruction by large scale parachute invasion or other such methods. Some of us might think that these factories have to be destroyed practically at *any price* but in order that *you* should be in a position to advise the War Cabinet how far to go in this respect you ought to have first-hand knowledge of the action radius within which life will be destroyed if a small atomic bomb is detonated above a city. Clearly you are the only member of the Cabinet who *can* have convictions based on his own computations rather than on official "reports" or other forms of "hearsay." If you check the calculations of others or make calculations of your own you will have a firm conviction of your own and will be able to arrive at a balanced recommendation and also be able to assist the War Cabinet in reaching what may be a very difficult decision.

If you wish to acquaint yourself with the calculations that have been made or can be made to determine the action radius of atomic bombs you will want to consider three different types of action which will destroy people, each having a different action radius.

The action radius of the blast that will destroy buildings as well as kill people is one of the three and you might find it rather difficult to arrive at a reliable estimate for this length, particularly since its value depends on such unknown factors as the quantity of material which is used for the construction of the German bomb and the manner in which it is detonated. Still such estimates give rise to grave concern if they are based on reasonable assumptions as to size and efficiency of the atomic bomb.

On the other hand, the action radius for killing people by radiation emitted at the moment of detonation from an atomic bomb which is exploded at a certain height above a city is practically independent of the construction and size of the bomb. You will therefore be able to obtain a fairly reliable estimate for this radius. You might then choose to consider an estimate for this radius as a lower limit for the action radius of *small bombs*.[13]

The third type of action is the cause of death by radiation due to the dispersal of the products *after* the detonation of the atomic bomb and might again be more difficult to estimate and perhaps impossible to express in the form of an action radius since meteorological factors may affect the result.[14]

I take it that any knowledge that may have been accumulated on this subject is accessible to you in the form of written reports, but from general experience I would not expect such reports to contain the pertinent information in an explicit form. Therefore you would perhaps wish to move faster and call in for consultation some of the British theoretical physicists who are up to date on this question. I do not know whether Peierls is up to date at present and whether he can be spared, but you may think of some other first class theoretical man whom you could use. It should not take too much time for a good man to acquaint himself with the facts if he is given the opportunity to have oral discussions with those members of your organization who are up to date on these questions.

I am certain you appreciate that I could not have made more specific statements in this letter without special authorization and to attempt to get that might have caused an undue delay. As it is I am informing the organization with which I am associated of the content of this personal letter and shall subsequently forward the letter by diplomatic mail.[15] My writing you may be a breach of etiquette from the official point

[13]In practice this radius is about 3/4 mile for a 20-kiloton Nagasaki-type bomb. At Hiroshima the 50% fatality radius was 0.8 mile, with more deaths due to blast and burns than to radiation. Samuel Glasstone, ed., *The Effects of Nuclear Weapons* (Washington: U.S. Atomic Energy Commission, 1962).
[14]This is an early mention of the "fallout" problem.
[15]The letter was forwarded August 25, after Bush gave his permission.

of view, but as I see it something more important than etiquette is at present involved. I feel confident that you will understand my reasons for writing this letter particularly if you decide to go more deeply into the study of the issues involved or have done so already.

With kind regards,

Yours sincerely,
Leo Szilard

The following memorandum[16] *was drafted for President Roosevelt, who did not live to see it; it was rewritten and placed before Byrnes. In it Szilard outlines far-reaching ideas on the implications of nuclear weapons and urges means to impose controls on their use—either by agreement with Russia or by accelerated production of great numbers of bombs. He also suggests the possibility of "denaturing" uranium and plutonium.*

Document 101

ATOMIC BOMBS AND THE POSTWAR POSITION OF THE UNITED STATES IN THE WORLD

Spring 1945

The development of the atomic bomb is mostly considered from the point of view of its possible use in the present war and such bombs are likely to be available in time to be used before the war ends. However, their role in the ten years which will follow can be expected to be far more important and it seems that the position of the United States in the world may be adversely affected by their existence. The following might very well turn out to be the future course of events:

Before the end of the war we shall use atomic bombs against Japan. These bombs will be much less powerful than we now know could be made and which in all likelihood will be made within two or three years; yet the first bomb that is detonated over Japan will be spectacular enough to start a race in *atomic* armaments between us and other nations.

In a few months Russia's war with Germany may be over. The work on uranium will then undoubtedly be given a high priority there but it will perhaps still not be carried out on a large industrial scale until we detonate our first atomic bomb and thus demonstrate the success of this development.[17] For a few years after that we shall almost certainly be ahead of Russia. But even if we assume that we could keep ahead of her in this development all the time, this may neither offer us protection from attack nor necessarily give us substantial advantage in case of war six years from now.

---

[16]Published with the omission of some then-classified technical data in the *Bulletin of the Atomic Scientists*, 5: 351–353 (December, 1947).

[17]This is apparently what happened. See Arnold Kramish, *Atomic Energy in the Soviet Union* (Stanford, Calif.: Stanford Univ. Press, 1959).

Six years from now Russia may have accumulated enough of some of the active elements which may be used for constructing atomic bombs to make atomic bombs which are equivalent to 10 million tons of TNT. Two tons of such active elements, if detonated with an efficiency of 30%, or ten tons of such elements, if detonated with an efficiency of 6%, would correspond to 10 million tons of TNT and this quantity would be sufficient to destroy all of our major cities in a single sudden attack.

Quoting the total amount of TNT to which an average atomic bomb corresponds does not give an adequate picture of the scope of action of such a bomb. A small bomb of this type corresponding to 10,000 tons of TNT detonated for instance at a suitable height above a city can be expected to destroy an area within a radius of one kilometer. A number of such bombs properly distributed over a city will make streets within a city completely impassable, may leave few survivors within the affected area, and can lead to total destruction by fire of the city.

A bomb containing about 200 lbs. of active material and weighing slightly more than a ton would, if detonated with an efficiency of 6%, correspond to 100,000 tons of TNT and destroy an area of about 4 square miles. The same bomb would, if detonated with 30% efficiency, destroy an area of 10 square miles.[18] Clearly, if such bombs are available, it is not necessary to bomb our cities from the air in order to destroy them. All that is necessary is to place a comparatively small number of such bombs in each of our major cities and to detonate them at some later time.

The United States has a very long coast-line which will make it possible to smuggle in such bombs in peacetime and to carry them by truck into our cities. The long coast-line, the structure of our society, and our very heterogeneous population may make an effective control of such "traffic" virtually impossible. One can easily visualize how a "friendly" power in time of peace may have such bombs placed in all of our major cities under the guidance of agents. This might be done free from aggressive intent. Such a power might know or suspect that we have accumulated a quantity of atomic bombs and that our defenses are so strong that after the outbreak of hostilities it would be difficult to reach our cities by air. In such circumstances it may be exceedingly difficult for its "government" to refuse to take "precautions" which its "army" considers necessary.

Such bombs may remain hidden in cellars of private houses in our cities for any number of years or they may remain hidden below the ground buried in gardens within our cities or buried in fields on the outskirts of our cities. Originally these bombs may have been planted merely as a routine precaution, but if later a serious international tension should develop there would be a strong temptation to exert pressure on the United States by virtue of the presence of these bombs. In case of war, all of our major cities might vanish within a few hours.

So far it has not been possible to devise any methods which would enable us to detect hidden atomic bombs buried in the ground or otherwise efficiently protected against detection.

[18] Less than 20 lbs. of plutonium is enough for a Nagasaki-type bomb (equivalent to 20,000 tons of TNT), but much larger fission bombs have been built since.

If there should be great progress in the development of rockets after this war it is conceivable that it will become possible to drop atomic bombs on the cities of the United States from very great distances by means of rockets.

The weakness of the position of the United States will largely be due to the very high concentration of its manufacturing capacity and of its *population* in cities. This concentration is so pronounced that the destruction of the cities may easily mean the end of our ability to resist. Keeping constantly ahead of the Russians in our production of these heavy elements will not restore us to a strong position. No quantity of these "active" materials which we may accumulate will protect us from attack and as far as retaliation is concerned we might not be able to do more than to destroy the large cities of Russia which are few in number and the economic importance of which is in no way comparable to the economic importance of our own cities. Thus it would appear that we would not gain an overwhelmingly strong position in a war with Russia merely by accumulating an enormous quantity of these elements or by increasing, as we might, the efficiency of our bombs from 6% to a much higher value.

The strong position of the United States in the world in the past thirty years was essentially due to the fact that the United States could out-produce every other country in heavy armaments. It takes a very large number of tanks, airplanes and guns to bring about a decision in a war and as long as tanks, airplanes and guns are the major instruments of war the large production capacity of the United States gives it an advantage which may be considered decisive.

The existence of atomic bombs means the end of the strong position of the United States in this respect. From now on the destructive power which can be accumulated by other countries as well as the United States can easily reach the level at which all the cities of the "enemy" can be destroyed in one single sudden attack. The expenditure in money and material which is necessary to reach this level is so small that any of the major powers can easily afford it provided they adopt "modern" production methods (see below). For us to accumulate active materials in quantities beyond that necessary to destroy the cities of the "enemy" would probably give some advantage in the war but it is difficult to say whether the importance of such "excess" amounts of material would be really substantial. Outproducing the "enemy" might therefore not necessarily increase our strength greatly.

The greatest danger arising out of a competition between the United States and Russia, which would lead to a rapid accumulation of vast quantities of atomic bombs in both countries, consists in the possibility of the outbreak of *a preventive war*. Such a war might be the outcome of the fear that the other country might strike first and no amount of good will on the part of both nations might be sufficient to prevent the outbreak of a war if such an explosive situation were allowed to develop.

One of the questions that has to be considered is whether it might be possible to set up some system of controls of the production of these active materials. Such controls would ultimately have to extend to every territory on the earth. Whether it is politically and technically feasible to set up effective controls and what we could do to improve our chances to bring this about are questions that urgently require study and decisions. Some further remarks on these questions are made below, but other considerations might be put forward as soon as the question receives the attention of the government.

The system of controls could be considered successful only if we could count on a period of grace in case the controls were denounced or obstructed by one of the major powers. This means that the system would have to be of such a nature that at least one or two years would lapse between the time the nations began to convert their installations for the purpose of manufacturing atomic bombs and the time such bombs became available in quantity.

The First Stage

Before going further it is necessary to make a few technical remarks:

The present development of the atomic bombs is based on methods which were devised in 1939 and 1940 and which must be considered as the *first stage* of the atomic power development. These methods are expensive in money and materials and most of them may be considered as out-dated.

This first stage may be defined by saying that it utilizes directly or indirectly only the energy locked up in the rare isotope of uranium.[19] Naturally found uranium contains less than one per cent of this rare isotope. It is doubtful whether the industrial installations based on this first stage will yield more than one ton of active material in the next couple of years which, taken with an efficiency of 6%, would correspond to about a million tons of TNT.

The first stage of this development is at present, so to speak, "in the bag". While in 1939 and 1940 the possibility of putting this first stage into operation was merely evidenced by the *assertions of the physicists*, the development has now reached the stage where the successful operation of this stage *can be demonstrated*, if need be, to skeptical statesmen.

The Second Stage

The second stage is characterized by the utilization of *the abundant isotope* (rather than the rare isotope) of uranium and would yield at a *low cost* vast quantities of the active materials.[20] With respect to this program we are today in a position similar to that which we occupied with respect to the first stage in 1940.

If conditions were created in which the physicists could work unhampered, it is

---

[19] This refers to separation of uranium-235 and to reactors powered by fission of uranium-235.
[20] Szilard was excited about the possibilities of fast-neutron "breeder" reactors. (Alvin Weinberg tells us that Szilard invented the term.) Breeders can convert uranium-238, the chief constituent of natural uranium, into plutonium, in large quantities. Szilard strongly advocated such reactors. They were later developed but are not as superior to slow-neutron reactors as he expected.

estimated that it would take *two years* to have this second stage in the pilot plant phase. Thirty tons of these active materials could be in production by means of these "modern" methods five years from now and the expenditure involved would be a small fraction of the cost which has so far been expended on the development of the now out-dated methods of the first stage.

It would appear highly desirable to set up at once an organization that is capable of taking care of the development of the second stage. But unless this development is coordinated with political action on the part of the United States Government, it will not materially contribute to the safety of the country; in certain unfavourable circumstances it may even be *detrimental* to the safety of the country.

Systems of Control Ought to be Considered

From a formal point of view all countries may be considered as potential enemies, but it is perhaps not too optimistic to assume that we may disregard the possibility of a war with Great Britain in the next fifteen years. It appears, however, rather unlikely that jointly with Great Britain we could police the world and thus prevent by force the manufacture of all of the "active materials" anywhere in the world including Russia.

It might perhaps be possible to set up jointly with Great Britain *and* Russia some sort of joint control of the manufacture of the active materials everywhere in the world if we could get Russia to agree to such a control *which of necessity would have to extend to her territory*. The purpose of such a control would be to prevent the active elements from becoming available in a form in which they could be used for the manufacture of atomic bombs. This does not necessarily mean that the development of atomic power is suppressed but only that the elements involved must not be prepared in certain forms and degree of purity.

This point raises the following question: What forms of atomic power can we permit to be organised if we want to make sure that the available materials and facilities cannot easily be converted for the manufacture of atomic bombs? Some thought has already been given to this question with the following results:

There are two types of active materials. Materials of the first type can be diluted by the abundant isotope of uranium in such a way as to rule out the possibility of using them for atomic bombs while leaving unimpaired the usefulness of the materials for industrial purposes.[21] A chemical separation from the diluting materials would be impossible and a conversion into materials which can be used for atomic bombs would take one or two years.

Material of the second type which can be used for atomic bombs can be "denatured" by adding a substance which cannot be separated chemically from it and

---

[21] Uranium-235 can be used in reactors even if it is diluted with enough uranium-238 to make it difficult or impossible to convert it directly into bomb material.

which will make it impossible to detonate by straightforward methods bombs which one may attempt to make from such mixtures. Whether more elaborate methods can be worked out which will permit the detonation of the denatured materials is a question which would have to carefully scrutinized.[22] These lines merely serve to indicate that there might perhaps be a satisfactory solution to the problem of reconciling the requirements of safety of the United States with the desire not to hamper the development of atomic power for industrial purposes.

Unfortunately it is by no means sure that a satisfactory solution of this problem is in fact possible. It would be much easier, safer, and would require a much less tight control to arrest the development of atomic power by scrapping and outlawing the large and easily visible installations which characterize *the first stage* of this development.

Control of Raw Materials Could be Considered

If Russia, the United States and other countries were willing to forego the use of atomic power for peacetime purposes one could have a system of control that would be fairly simple since it would be sufficient essentially to control the movements of raw materials. Ores of uranium would have to be mined under control and transported to some "neutral" territory. Whether or not it would be permitted to have in a neutral territory installations belonging to the first stage and atomic power plants would be a question of minor importance. It is likely that if the major powers were willing to forego the use of atomic power it would seem that a system of controls could be set up without encountering too great difficulties.

*An Alternative System of
Controls Would Have To be Much Tighter*

On the other hand, if the United States, Russia, and other countries should have atomic power installations within their own territory, a very tight system of control would be needed in order to make sure that the United States would not have to face a sudden attack by atomic bombs. For a control of this sort to be effective, it would be necessary that our agents and the agents of Great Britain move freely around in Russia, be permitted to keep contacts with Russian civilians, secretly employ Russian civilians for the purpose of obtaining information, and have entry into every factory or shop throughout the vast territory of Russia.

That there may be dangerous loopholes in control systems which might be set up is illustrated by events that took place in Germany after the first World War. At that time there were many Germans who were willing to give information to the Inter-

---

[22]The "denaturing" of plutonium with plutonium-240 (which emits particles that tend to make a mass explode prematurely) is not in fact foolproof.

Allied Commission about violations of the Control regulations, but those who actually did so were publicly tried under the German Espionage Law and were given heavy sentences. The Treaty of Versailles did not stipulate that the German Espionage Law must be revoked.

Clearly, it would be desirable to create a situation which would permit us to appeal in various ways to physicists and engineers everywhere for information that would uncover violations of the controls. This would give us additional asurance that such violations would be detected but it presupposes that we succeed in creating conditions that would enable us to guarantee the personal safety of those who volunteer such information and the safety of their families.

Since Russia cannot be expected to agree to such a control unless she obtain the same rights of control in the United States and Great Britain the question whether Congress and the people of the United States are willing to agree to such a control might become of paramount importance.

How Could Russia Best be Persuaded?

As to our chances of persuading the Russians to accept mutual control, much may depend on the proper timing of our approach to Russia. It would appear that such an approach would have to be made immediately after we demonstrate the potency of atomic bombs.

Such a demonstration may take place in the course of the war. However, the psychological advantages of avoiding the use of atomic bombs against Japan and, instead, of staging a demonstration of the atomic bomb at a time which appears most appropriate from the point of view of its effect on the governments concerned might be very great. Therefore this possibility seems to deserve serious consideration in deciding whether or not to use such a bomb against Japan. If at the time when we demonstrate the atomic bomb to the world we had the second stage of the atomic power development "in the bag", chances of obtaining the consent of Russia to some satisfactory system of controls might be considerably improved. At that time, Russian physicists would probably be quite uncertain as to whether or not they could catch up with us in this development. As far as the first stage is concerned, they may be expected to have a full appreciation of its scope and to be fairly confident that they can duplicate in a fairly short time what we have accomplished so far. Their knowledge of atomic constants, however, is in all probability too inaccurate to enable them to appraise whether or not they can utilize the abundant isotope of uranium or to estimate how long it would take them and how much it would cost. As long as the Russian physicists remain uncertain about this point there might be considerable willingness on the part of the Russian government to set up jointly with us and Great Britain a really effective control of this field. Whether or not we have by that time actually accumulated ten or twenty tons of the active substances appears to be of secondary importance as long as we can demonstrate that we have in manufacture quantities which cannot be derived *from the first stage of development* with which the Russians may be familiar.

Events may be expected to move so fast that if it is intended to reach an agreement with Russia and other countries such an agreement would have to be complete before the next presidential elections. Thus we have to conclude that if the second stage is to be developed for the purpose of enabling the Administration to obtain Russian cooperation in the proposed joint control, we have no time to lose in attacking the problems connected with the second stage of the development.

If the Control is Interfered with

While it may be a great step forward to establish a tight control on the atomic power development by a reciprocal agreement with Great Britain and Russia and extend it to all territories of the world, yet we cannot disregard the possibility that one of the major powers, for instance Russia, after a few years—during which the controls may have operated quite successfully—may begin to place difficulties in the way of an effective control of activities conducted in its own territory. Clearly it would be quite essential that the people of this country and the world be brought to understand from the start that any difficulties which any nation may place in the way of the established controls would have to be considered as tantamount to a "declaration of war".

Such a "declaration of war" would have the effect that the United States and other countries involved would at once begin to manufacture atomic bombs. If up to that time the control had been effective, it would take about two years to convert the materials and installations involved in the utilization of atomic power to the manufacture of bombs. In such an "armament race" in which all countries would have to start, so to speak, from scratch, the position of the United States might be quite favorable, provided the development of atomic power had been kept up at high level.

Clearly if any major power deliberately wants to start a war, there will be a war and all that we can hope to achieve by the reciprocal control which we have discussed is that a war may not break out *as a result of an armament race.*

Still, it would seem that if the situation were generally understood there might be some hope that having succeeded in setting up a system of reciprocal control and having kept it in operation for a few years, neither the United States nor Great Britain nor Russia would attempt to interfere with this system of control in such a manner that its acts would be considered by the other partners as a menace to their security. We would then perhaps have a chance of living through this century without having our cities destroyed.

An attempt to manufacture atomic bombs undertaken by any of the smaller countries would be of minor importance since it could be met by immediate armed intervention using ordinary methods of warfare such as tanks and airplanes.

In the Absence of a System of Controls

In discussing our postwar situation the greatest attention was given in this memorandum to the role that Russia might play. This was not done because it was assumed that Russia may have aggressive intentions but rather because it was assumed that if an agreement can be reached with Russia it will be possible to extend the system of controls to every country in the world.

In the absence of a system of controls, however, a number of countries might, say, ten years from now, be in possession of large quantities of atomic bombs and represent a threat to the cities of the United States.

What policies could be adopted to safeguard the security of the United States in the absence of a reliable system of controls requires serious consideration particularly since our chances of creating a satisfactory system of controls may be rather small. The situation will probably have some effect on city planning.

In discussing this question one will have to consider a number of possibilities which go far beyond the narrower question of whether or not the second stage of the atomic power development ought to be vigorously pursued, and the discussion of those possibilities goes beyond the scope of the present memorandum. As far as this narrower question is concerned the following remark might, however, be made.

One might consider the advisability of discontinuing now the work on detonating active substances and of immediately scrapping now all installations for the manufacture of active materials. In view of the fact that the Germans have not pushed this development, the scrapping of our own installations coupled with an agreement with Russia and Great Britain which would outlaw the building of such installations might perhaps enhance the security of the United States in the next 25 years. In order to understand this point of view one has to realize that it is necessary to develop the first stage of atomic power before the *second stage* can be entered upon[23] and that the installations belonging to the first stage are of necessity large and conspicuous structures; consequently it *does not require a tight control* to detect any structures of this type which might be erected in violation of the law.

Conversely, it might be proposed that we should *lose no time* in developing the second stage of atomic power and that we should develop within a few years methods for manufacturing overwhelming quantities of the active materials.

While it may be difficult to decide between these two points of view, the present trend to develop atomic bombs and to maintain our installations for the manufacture of active materials but to delay in developing the *second stage* would appear to lead to the worst possible course of action that we could take.

The following letters trace the attempts to bring Szilard's ideas before first Roosevelt and then Truman:

---

[23]Since uranium-235 or plutonium is needed to fuel the first breeder reactor.

Document 102

112 Mercer Street
Princeton, New Jersey
March 25, 1945

The Honorable Franklin Delano Roosevelt
The President of the United States
The White House
Washington, D.C.

Sir:

I am writing you to introduce Dr. L. Szilard who proposes to submit to you certain considerations and recommendations. Unusual circumstances which I shall describe further below induce me to take this action in spite of the fact that I do not know the substance of the considerations and recommendations which Dr. Szilard proposes to submit to you.

In the summer of 1939 Dr. Szilard put before me his views concerning the potential importance of uranium for national defense. He was greatly disturbed by the potentialities involved and anxious that the United States Government be advised of them as soon as possible. Dr. Szilard, who is one of the discoverers of the neutron emission of uranium on which all present work on uranium is based, described to me a specific system which he devised and which he thought would make it possible to set up a chain reaction in unseparated uranium in the immediate future. Having known him for over twenty years both from his scientific work and personally, I have much confidence in his judgment and it was on the basis of his judgment as well as my own that I took liberty to approach you in connection with this subject. You responded to my letter dated August 2, 1939 by the appointment of a committee under the chairmanship of Dr. Briggs and thus started the Government's activity in this field.

The terms of secrecy under which Dr. Szilard is working at present do not permit him to give me information about his work; however, I understand that he now is greatly concerned about the lack of adequate contact between scientists who are doing this work and those members of your Cabinet who are responsible for formulating policy. In the circumstances I consider it my duty to give Dr. Szilard this introduction and I wish to express the hope that you will be able to give his presentation of the case your personal attention.[24]

Very truly yours,
A. Einstein

[Enclosure to Einstein's letter by Szilard][25]

The work on uranium has now reached a stage which will make it possible for the

[24]A copy of this letter is in the Szilard files. It was also published in *Einstein on Peace*, pp. 304–305. Reprinted by permission of the Estate of Albert Einstein.
[25]This original version was modified after Roosevelt's death, then given to Byrnes for transmittal to President Truman.

Army to detonate atomic bombs in the immediate future. The "demonstration" of such bombs may be expected rather soon and naturally the War Department is considering the use of such bombs in the war against Japan.

From a purely military point of view this may be a favorable development. However, many of those scientists who are in a position to make allowances for the future development of this field believe that we are at present moving along a road leading to the destruction of the strong position that the United States hitherto occupied in the world. It appears probable that it will take just a few years before this will become manifest.

Perhaps the greatest immediate danger which faces us is the probability that our "demonstration" of atomic bombs will precipitate a race in the production of these devices between the United States and Russia and that if we continue to pursue the present course, our initial advatage may be lost very quickly in such a race.

If a nation were to start now to develop atomic bombs, so to speak from scratch, it could do so without reproducing many of the expensive installations which were built by the War Department during the War. *For over a year now we have known that we could develop methods by means of which atomic bombs can be produced from the main component of uranium which is more than one hundred times as abundant as the rare component* from which we are manufacturing atomic bombs at present. We must expect that [at] a cost of about $500 million some nations may accumulate, within six years, a quantity of atomic bombs that will correspond to ten million tons of TNT. A single bomb of this type weighing about one ton and containing less than 200 pounds of active material may be expected to destroy an area of ten square miles. Under the conditions expected to prevail six years from now, most of our major cities might be completely destroyed in one single sudden attack and their populations might perish.

In the United States, thirty million people live in cities with a population of over 250,000 and a consideration of this and other factors involved indicates that the United States will be much more vulnerable than most other countries.

Thus the Government of the United States is at present faced with the necessity of arriving at decisions which will control the course that is to be followed from here on. These decisions ought to be based not on the *present* evidence relating to atomic bombs, but rather on the situation which can be expected to confront us in this respect a few years from now. This situation can be evaluated only by men who have firsthand knowledge of the facts involved, that is, by the small group of scientists who are actively engaged in this work. This group includes a number of eminent scientists who are willing to present their views; there is, however, no mechanism through which direct contact could be maintained between them and those men who are, by virtue of their position, responsible for formulating the policy which the United States might pursue.

The points on which decisions appear to be most urgently needed are as follows:

1. Shall we aim at trying to avoid a race in the production of atomic bombs between the United States and certain other nations?

2. Can a system of controls relating to this field be devised which is sufficiently tight to be relied on by the United States and which had some chance of being accepted under otherwise favorable conditions by Russia and Great Britain?

3. Can we materially improve our chances to obtain the cooperation of Russia in setting up such a system of controls by developing in the next two years modern methods of production which would give us an overwhelming superiority in this field at the time when Russian might be approached?

4. What framework could immediately be set up within which the scientific development of such "modern" methods could vigorously be pursued both under present and postwar conditions? Should, for instance, this framework be set up under the Secretary of Commerce or under the Secretary of the Interior, or should the scientific development be under a Government-owned corporation jointly controlled by the Secretary of Commerce, the Secretary of the Interior, and the Secretary of War?

5. Should the scientific development work be based on the assumption that a race in the production of atomic bombs is unavoidable and accordingly be aimed at maximum potential of war, say in six years from now, or should the scientific development be rather aimed at putting us into a favorable position with respect to negotiations with our Allies two or three years from now?

6. Should, in the light of the decisions concerning the above points, our "demonstration" of atomic bombs and their use against Japan be delayed until a certain further stage in the political and technical development has been reached so that the United States shall be in a more favorable position in negotiations aimed at setting up a system of controls?

Other decisions which are needed but which are perhaps less urgent, would come within the competence of the Department of the Interior.

If there were in existence a small subcommittee of the Cabinet (having as its members the Secretary of War, either the Secretary of Commerce or the Secretary of the Interior, a representative of the State Department, and a representative of the President, acting as the secretary of the Committee), the scientists could submit to such a committee theirr ecommendations either by appearing from time to time before the committee or through the secretary of the committee.

The latter, if so authorized by the President, could also act as a liaison to the scientists prior to the designation of such a subcommittee. At his disposal could then be placed a memorandum which had been prepared in an attempt to analyze the consequences of the scientific and technical development which we have to anticipate. The memorandum was prepared on the basis of consultations with ten scientists from six different institutions in the United States. These and other eminent scientists who were not consulted would undoubtedly avail themselves of the opportunity of presenting their views to a man authorized by the President, assuming that such a man would have the time at his disposal which a study of this kind would require.

Document 103

The Quadrangle Club
Chicago
115 East 57th St.
May 25, 1945

The Honorable Harry S. Truman
The President of the United States
The White House
Washington, D.C.

Sir:

I have the honor to transmit to you a letter of introduction written by Professor Albert Einstein to the late President of the United States to whom—on account of his early death—I was unable to present it. The document to which Mr. Einstein's letter refers is attached as a second inclosure and is respectfully submitted for your consideration.

Very truly yours,
Leo Szilard

Szilard now considered drafting a petition of scientists to the President. Meanwhile, as the following letter indicates, he was vigorously debating the issues with his colleagues.

Document 104

[Teller]
P.O. Box 1663
Santa Fe, New Mexico[26]
July 2, 1945

Dr. Leo Szilard
P.O. Box 5207
Chicago 80, Illinois

Dear Szilard:

Since our discussion I have spent some time thinking about your objections to an immediate military use of the weapon we may produce. I decided to do nothing. I should like to tell you my reasons.

First of all let me say that I have no hope of clearing my conscience. The things we are working on are so terrible that no amount of protesting or fiddling with politics will save our souls.

This much is true: I have not worked on the project for a very selfish reason and I have gotten much more trouble than pleasure out of it. I worked because the problems interested me and I should have felt it a great restraint not to go ahead. I can not claim that I simply worked to do my duty. A sense of duty could keep me out of such

---

[26]Address for the Los Alamos nuclear weapons laboratory.

work. It could not get me into the present kind of activity against my inclinations. If you should succeed in convincing me that your moral objections are valid, I should quit working. I hardly think that I should start protesting.

But I am not really convinced of your objections. I do not feel that there is any chance to outlaw any one weapon. If we have a slim chance of survival, it lies in the possibility to get rid of wars. The more decisive a weapon is the more surely it will be used in any real conflict and no agreements will help.

Our only hope is in getting the facts of our results before the people. This might help to convince everybody that the next war would be fatal. For this purpose actual combat-use might even be the best thing.

And this brings me to the main point. The accident that we worked out this dreadful thing should not give us the responsibility of having a voice in how it is to be used. This responsibility must in the end be shifted to the people as a whole and that can be done only by making the facts known. This is the only cause for which I feel entitled in doing something: the necessity of lifting the secrecy at least as far as the broad issues of our work are concerned. My understanding is that this will be done as soon as the military situation permits it.

All this may seem to you quite wrong. I should be glad if you showed this letter to Eugene[27] and Franck who seem to agree with you rather than with me. I should like to have the advice of all of you whether you think it is a crime to continue to work. But I feel that I should do the wrong thing if I tried to say how to tie the little toe of the ghost to the bottle from which we just helped it to escape.[28]

With best regards,

Yours,
E. Teller

Szilard wrote several drafts[29] of a petition (similar to the final draft printed below, except for its conclusion) and circulated it among the scientists at Chicago. Here is the covering letter for an early draft.

Document 105

July 4, 1945

Dear—:

Inclosed is the text of a petition which will be submitted to the President of the United States. As you will see, this petition is based on purely moral considerations.

---

[27] Probably Eugene Wigner.
[28] Teller later wrote that he was "in absolute agreement" with the sentiments of Szilard's petition, but that J. R. Oppenheimer's arguments changed his mind. Teller with Allen Brown, *The Legacy of Hiroshima* (Garden City, N.Y.: Doubleday, 1962), pp. 13–14. Cf. Martin J. Sherwin, *A World Destroyed. The Atomic Bomb and the Grand Alliance* (New York: Knopf, 1975), pp. 218–219.
[29] The draft of July 3 is printed as Appendix IV of Szilard's "Reminiscences," *Perspectives in American History*, 2: 94–151 (1968).

It may very well be that the decision of the President whether or not to use atomic bombs in the war against Japan will largely be based on considerations of expediency. On the basis of expediency, many arguments could be put forward both for and against our use of atomic bombs against Japan. Such arguments could be considered only within the framework of a thorough analysis of the situation which will face the United States after this war and it was felt that no useful purpose would be served by considering arguments of expediency in a short petition.

However small the chance might be that our petition may influence the course of events, I personally feel that it would be a matter of importance if a large number of scientists who have worked in this field went clearly and unmistakably on record as to their opposition on moral grounds to the use of these bombs in the present phase of the war.

Many of us are inclined to say that individual Germans share the guilt for the acts which Germany committed during this war because they did not raise their voices in protest against those acts. Their defense that their protest would have been of no avail hardly seems acceptable even though these Germans could not have protested without running risks to life and liberty. We are in a position to raise our voices without incurring any such risks even though we might incur the displeasure of some of those who are at present in charge of controlling the work on "atomic power."

The fact that the people of the United States are unaware of the choice which faces us increases our responsibility in this matter since those who have worked on "atomic power" represent a sample of the population and they alone are in a position to form an opinion and declare their stand.

Anyone who might wish to go on record by signing the petition ought to have an opportunity to do so and, therefore, it would be appreciated if you could give every member of your group an opportunity for signing.

The draft mentioned in the foregoing concluded by petitioning the President "to rule that the United States shall not, in the present phase of the war, resort to the use of atomic bombs." Eighteen people signed the following revision.

Document 106

July 13, 1945

We, the undersigned, agree in essence with the attached petition, but feel that our attitude is more clearly expressed if its last paragraph is replaced by the following:

We respectfully petition that the use of atomic bombs, particularly against cities, be sanctioned by you as Chief Executive only under the following conditions:
 1. Opportunity has been given to the Japanese to surrender on terms assuring them the possibility of peaceful development in their homeland.
 2. Convincing warnings have been given that a refusal to surrender will be followed by the use of a new weapon.
 3. Responsibility for use of atomic bombs is shared with our allies.

Szilard now wrote a final draft and circulated it in the Metallurgical Laboratory. The original copy of this petition (returned to Szilard in 1957) bears 68 signatures.

Document 107

A PETITION TO THE PRESIDENT OF THE UNITED STATES
July 17, 1945

Discoveries of which the people of the United States are not aware may affect the welfare of this nation in the near future. The liberation of atomic power which has been achieved places atomic bombs in the hands of the Army. It places in your hands, as Commander-in-Chief, the fateful decision whether or not to sanction the use of such bombs in the present phase of the war against Japan.

We, the undersigned scientists, have been working in the field of atomic power. Until recently we have had to fear that the United States might be attacked by atomic bombs during this war and that her only defense might lie in a counterattack by the same means. Today, with the defeat of Germany, this danger is averted and we feel impelled to say what follows:

The war has to be brought speedily to a successful conclusion and attacks by atomic bombs may very well be an effective method of warfare. We feel, however, that such attacks on Japan could not be justified, at least not until the terms which will be imposed after the war on Japan were made public in detail and Japan were given an opportunity to surrender.

If such public announcement gave assurance to the Japanese that they could look forward to a life devoted to peaceful pursuits in their homeland and if Japan still refused to surrender our nation might then, in certain circumstances, find itself forced to resort to the use of atomic bombs. Such a step, however, ought not to be made at any time without seriously considering the moral responsibilities which are involved.

The development of atomic power will provide the nations with new means of destruction. The atomic bombs at our disposal represent only the first step in this direction, and there is almost no limit to the destructive power which will become available in the course of their future development. Thus a nation which sets the precedent of using these newly liberated forces of nature for purposes of destruction may have to bear the responsibility of opening the door to an era of devastation on an unimaginable scale.

If after the war a situation is allowed to develop in the world which permits rival powers to be in uncontrolled possession of these new means of destruction, the cities of the United States as well as the cities of other nations will be in continuous danger of sudden annihilation. All the resources of the United States, moral and material, may have to be mobilized to prevent the advent of such a world situation. Its prevention is at present the solemn responsibility of the United States—singled out by virtue of her lead in the field of atomic power.

The added material strength which this lead gives to the United States brings with it the obligation of restraint and if we were to violate this obligation our moral position would be weakened in the eyes of the world and in our own eyes. It would then be more difficult for us to live up to our responsibility of bringing the unloosened forces of destruction under control.

In view of the foregoing, we, the undersigned, respectfully petition: first, that you exercise your power as Commander-in-Chief to rule that the United States shall not resort to the use of atomic bombs in this war unless the terms which will be imposed upon Japan have been made public in detail and Japan knowing these terms has refused to surrender; second, that in such an event the question whether or not to use atomic bombs be decided by you in the light of the consideration presented in this petition as well as all the other moral responsibilities which are involved.

The following is an unsigned copy of a letter by Szilard, documenting how the first draft petition was sent to the nuclear bomb laboratory at Los Alamos:

Document 108

Metallurgical Laboratory
July 10, 1945

Dr. Ed Creutz[30]
Box 1633
Los Alamos, N.M.

Dear Creutz:

I have sent to you with Lapp[31] eight sets of a "petition" which has been circulating in Chicago for some time.

Please give one set to Oppenheimer[32] for his information and give the other sets to such men as are willing to circulate them. You may address one set to Teller, one to Bob Wilson, one to yourself, one to Morrison, one to McMillan,[33] etc., but please use your own judgment and ask the persons involved if they are willing to circulate it in their group.

Before circulating this petition you had better stamp all copies "secret". It would be desirable to have the circulation finished in five days from the receipt of the petition and I will communicate with you within one week in order to discuss where the signed copies should be sent.

Of course, you will find only a few people on your project who are willing to sign such a petition and I am sure you will find many boys confused as to what kind of a thing a moral issue is.

[30]Edward Creutz, physicist at Los Alamos.
[31]Ralph Lapp, assistant to Compton at the Metallurgical Laboratory.
[32]This may refer to Frank Oppenheimer, or more likely to his brother J. Robert (see n. 34, below). Both of them were physicists at Los Alamos.
[33]Robert Wilson, Philip Morrison, and Edwin McMillan were also physicists at Los Alamos.

Clearly, there are more important things to do and more effective ways to influence the course of events than the circulating of a petition but I have no doubt in my own mind that from a point of view of the standing of the scientists in the eyes of the general public one or two years from now it is a good thing that a minority of scientists should have gone on record in favor of giving greater weight to moral arguments and should have exercised their right given to them by the Constitution to petition the President. With kind regards,

Very sincerely yours,

P.S. Please give the enclosed letter to Oppenheimer[34] together with a set of the petition.

The petition sent to Los Alamos was not circulated, apparently at Oppenheimer's direction.[35] Szilard also tried (without success) to get signatures at Oak Ridge and to have the signed Chicago copy forwarded to the President.

Document 109

July 7, 1945

Mr. M. D. Whitaker
P.O. Box 1991
Knoxville 11, Tenn.
Attention: Mr. E. P. Wigner

Dear Wigner:

Enclosed you will find a number of copies of the petition, two of each attached to one covering letter. I have signed the covering letters but I leave it to you to put in on top of the covering letter the name of the group or section leader to whom you want to hand it.

Please keep a list of those to whom you gave copies of the petition and draw their attention to the fact that this petition is a secret document. The petition ought to be mailed back to me no later than Saturday, July 14th and preferably they ought to be mailed back Thursday, the 12th. No one ought to sign the petition who does not feel that he wants to go on record in this matter.

Sincerely,
Leo Szilard

LS: jjp
encls.

P.S. No member of the armed services ought to be asked to sign this petition.

---

[34]This letter is addressed to F. Oppenheimer; it is similar to the letter to Creutz, but briefer.
[35]i.e., J. Robert Oppenheimer, director of Los Alamos. See Edward Teller with Allen Brown, *The Legacy of Hiroshima*, 1962, pp. 13–14.

Document 110

Metallurgical Laboratory
P.O. Box 5207
Chicago 80, Illinois
July 19, 1945

Dr. A. H. Compton
Metallurgical Laboratory
Chicago, Illinois

Dear Dr. Compton:

Enclosed is a petition to the President of the United States signed by 67 scientists residing in Chicago. You were good enough to say that you would forward this petition to the President via the War Department. Since those who have signed this petition are exercising a privilege under the Constitution, I believe that we are not free to disclose their names to anyone but to those who are authorized to open the mail of the President. May I, therefore, suggest that the petition be placed in an envelope addressed to the President and that the envelope be sealed before it leaves your office.

Enclosed are six unsigned copies of the petition which you may wish to communicate to others who ought in your opinion to be informed of the text of the petition.

It would be appreciated if in transmitting these copies you would draw attention in your covering letter to the fact that the text of the petition deals with the moral aspect of the issue only. Some of those who signed the petition undoubtedly fear that the use of atomic bombs at this time would precipitate an armament race with Russia and believe that atomic bombs ought not be demonstrated until the government had more time to reach a final decision as to which course it intends to follow in the years following the first demonstration of atomic bombs. Others are more inclined to think that if we withhold such a demonstration we will cause distrust on the part of other nations and are, therefore, in favor of an early demonstration. The text of the petition does not touch upon these and other important issues involved but deals with the moral issue only.

Very sincerely yours,
Leo Szilard

jjp
encls.

Document 111

July 24, 1945

To: Colonel K. D. Nichols
From: Arthur H. Compton
In re: Transmittal of Petitions addressed to the President

I have been requested to transmit the enclosed petition to the President of the United States. At the suggestion of General Groves, I am herewith handing it to you for disposition. Since the matter presented in the petition is of immediate concern, the petitioners desire the transmittal to occur as promptly as possible. It will be appreciated if you will inform me with regard to its disposition.

You will note that the signed draft of the petition is enclosed within a sealed envelope. I have personally verified that this envelope contains only signed copies of a petition, identical in text with the carbon copy attached, together with receipt forms for classified material. Mr. Szilard, in his covering letter, has requested that this envelope be opened only by those authorized to read the President's mail.

You have requested me to evaluate this petition and likewise those submitted to you by Mr. Whitaker[36] on behalf of certain members of Clinton Laboratories.

The question of use of atomic weapons has been considered by the Scientific Panel of the Secretary of War's Interim Advisory Committee. The opinion which they expressed was that military use of such weapons should be made in the Japanese War. There was not sufficient agreement among the members of the panel to unite upon a statement as to how or under what conditions such use was to be made.

A small group of petitioners initially canvassed certain groups of scientists within the project seeking signatures requesting no use of the new weapons in this war. The response was such as to call forth several counter petitions, of which those submitted through Mr. Whitaker are typical, and to cause the formulator of the original petition to rephrase it so as to approve use of the weapons after giving suitable warning and opportunity for surrender under known conditions.

In order to obtain a fair expression of the opinion of a typical group of scientists, an opinion poll was conducted on a group of 150. The results are described in the enclosed memo to me from Dr. Daniels.[37] You will note that the strongly favored procedure is to "give a military demonstration in Japan, to be followed by a renewed opportunity for surrender before full use of the weapons is employed." This coincides with my own preference, and is, as nearly as I can judge, the procedure that has found most favor in all informal groups where the subject has been discussed.

After the destruction of Hiroshima, Szilard tried to declassify the petition so he could publish it:

Document 112

Metallurgical Laboratory
P.O. Box 5207
Chicago 80, Illinois
August 17, 1945

Mr. Matthew J. Connelly
The White House
Washington, D.C.

Dear Mr. Connelly:

When Mr. Bartky and I called on you on May 25, you were kind enough to arrange

---

[36] Martin D. Whitaker, director of the Clinton Engineer Works, Oak Ridge, Tenn.
[37] Farrington Daniels, director of the Metallurgical Laboratory. See Alice K. Smith, *A Peril and A Hope: The Scientists' Movement in America, 1945–1947* (Chicago: University of Chicago Press, 1965), pp. 57–59; hereinafter referred to as Alice Smith, *A Peril and A Hope*.

an interview with Mr. Byrnes. H. C. Urey of Columbia University, Walter Bartky of the University of Chicago, and I saw Byrnes on May 28 and submitted to him a memorandum dated Spring, 1945 which was originally prepared for Mr. Roosevelt and which you have read. We are very grateful to you for the opportunity to present our views to Mr. Byrnes.

The enclosed envelope contains Mr. Einstein's letter, returned by Mr. Byrnes for transmittal to your office, and a copy of the memorandum which we left with Mr. Byrnes. You had previously seen both of these documents and they are merely transmitted for your files.

Enclosed also is the text of a petition which was signed by 67 scientists working in this Laboratory. It may not have crossed your desk since it had been transmitted in July via the War Department. Some of those who signed this petition have asked me that its text be now made public; and I wondered whether you would be good enough to let me know by August 24 if you considered its publication undesirable.

Very sincerely yours,
Leo Szilard

Document 113

THE WHITE HOUSE WASHINGTON DC 1945 AUG 25

DR LEO SZILARD
UNIVERSITY OF CHICAGO

REFERENCE YOUR TELEPHONE CALL THE PRESIDENT HAS YOUR LETTER UNDER ADVISEMENT. I WILL BE GLAD LET YOU KNOW HIS DECISION LATER. REGARDS

MATTHEW J CONNELLY SECRETARY TO THE PRESIDENT

Document 114

Armed Service Forces
Manhattan Engineer District
Intelligence and Security Division
Chicago Branch Office
P. O. Box 6770-A
Chicago 80, Illinois
27 August 1945

Dear Dr. Szilard:

Pursuant to our telephone conversation on 25 August 1945, I am submitting this letter to you to set forth, in writing, the reasons for my oral request that you reclassify the petition to the President of the United States dated 17 July 1945, of which you are the author.

Primarily, for purposes of review, I want to outline briefly certain discussions which have occurred between the Military Intelligence Division and yourself in connection with the petition and its military classification:

a. It is understood that when this petition was originally drawn you did not assign a military classification to it.

b. Sometime subsequent to the date of your petition you were informed by Major C. C. Pierce of the Washington Liaison Office of the Manhattan District that the petition should bear a military classification of "Secret". You agreed as to the justification for such a classification and it was so classified.

c. On 11 August 1945, you directed a letter to Captain J. H. McKinley stating that the petition "will no longer be treated as a classified document." You informed me on 16 August 1945 of this letter to Captain McKinley and I told you that the petition could now be declassified. My authority to so advise you was based on permission which I had received from my superiors in this division.

d. Subsequently, on 25 August, I was telephonically advised by my superiors that the question of a military classification for your petition had been reviewed by Major General L. R. Groves and that he, in the light of certain statements in the petition, as well as the very nature of the petition itself when coupled with certain world developments having military significance, had determined in the exercise of his best judgment to request that the petition be again classified secret by you with its attendant limitations. You were given this information telephonically on 25 August. You then requested a written statement officially placing our request for reclassification before you.

By definition, the military classification of "Secret" includes: "Information, or features contained therein, the disclosure of which might endanger national security, cause serious injury to the interest or prestige of the nation or any governmental activity, or would be of great advantage to a foreign nation. . . . " Army Regulations 380-5, dated 15 May 1944; see also Intelligence Bulletin Number Five, Manhattan Engineer District, revised 1 Sept. 1944.

The authority to place a military classification of secret on documents is rather severely limited by the War Department. Civilians, normally, do not possess this authority. However, in the Manhattan District, such authority has upon occasion been delegated by the military authorities to the heads of organizations working for the Manhattan District and to certain other civilians designated by these heads. It is assumed that you have been one of those designated by Dr. Compton.

The authority to classify implies the authority to refrain from classifying, that is, to decide whether or not a certain document should bear any classification. Granted that you have the authority to classify or not to classify certain documents, any such authority which you possess is a delegated authority stemming from General Groves himself, through Dr. A. H. Compton, to you. It is, like all delegated authority in our government, subject to review by the delegator. General Groves has so reviewed your petition and your decision to declassify it and has determined, in the light of what must be conceded is a wider knowledge of the scope and present ramifications of the atomic bomb program, that the petition should be classified secret and hence its dissemination must be appropriately limited.

The knowledge which you have acquired by virtue of your position as an employee of the Metallurgical Project of the University of Chicago which, in turn, is supervised by the Manhattan District, has been, it must be assumed, the basis upon which you wrote your petition. The petition predicates a knowledge of the scope, objectives and potentialities of the Manhattan District Project, information concerning which you acquired by virtue of your official position with the Project.

You will recall that on 25 February 1942 you solemnly swore to "not by any means divulge or disclose any secret or confidential information" that you might obtain or acquire by reason of your connection with the N.D.R.C. unless authorized to do so.[38] Since that N.D.R.C. work meshed into the O.S.R.D.[39] and it, in turn, into the Manhattan District, I believe that any lawyer would advise you that secret and confidential information you acquired from your connection with the Manhattan District would come within the purview of this promise by you.

You have, from time to time, signed certain other secrecy agreements, Espionage Act declarations, and patent agreements with the United States as well as committments in your present contract of employment and its supplement, all of which preclude the disclosure by you of any information considered secret by the head of the Manhattan District, Major General L. R. Groves.

It appears unnecessary to point out to you that any information considered "secret" by the highest authority which you divulge to persons unauthorized to receive it will be in violation of the above agreements and of the Espionage Act (Title I, Sec. 1, 40 Stat. 217 as amended by Pub. Act No. 443, Laws of 1940) and for which you may be held strictly accountable.

Every effort is being made by General Groves and those above him to authorize the release of all information concerning the project which can be released without jeopardizing the safety or welfare of the people of the United States. It was the considered opinion of General Groves and those above him that your petition did not fall within the purview of such information which could be released without jeopardy.

You asked me to point out certain passages in your petition which might be considered as justification for General Groves' belief that it should be classified secret. The opinions which I give you are my own and the ones I would use in determining whether or not the petition should be classified. In my opinion, then, every paragraph of the petition either contains some information or implies "inside" information, i.e., information gained through employment, which, when linked with the purpose of the petition, implies that internal dissension and fundamental differences in point of view disrupted the development and fruition of the District's work—an implication which you as well as I know is not founded on sober fact and which, if released at this time,

---

[38] Such affidavits were routinely required of everyone doing secret work for the National Defense Research Committee (N.D.R.C.).
[39] Office of Scientific Research and Development.

might well cause "injury to the interest or prestige of the nation or governmental activity." Therefore, it follows that, in my opinion, the entire petition should be classified secret with no exception for any one paragraph.

As you already know, the War Department has set up a proper channel through which information can be released to the press or classified information can be brought to the attention of those government officials charged with determining the future of the Project. The first channel is through Lt. Col. W. A. Consodine, P. O. Box 2610, Washington, 25, D.C., and the second, in your case, is through Dr. A. H. Compton.

Sincerely,
James S. Murray
Captain, Corps of Engineers
Intelligence Officer

cc:

Lt. Col. W. B. Parsons, P. O. Box "W", Oak Ridge, Tenn.
Major Claude C. Pierce, Jr., P.O. Box 2610, Washington, D.C.
Capt. J. H. McKinley, P.O. Box 6140-A, Chicago 80, Ill.

Next Szilard was warned (not for the first time) that he could lose his job if he disobeyed the Army's rules:

Document 115

Armed Service Forces
Manhattan Engineer District
Intelligence and Security Division
Chicago Branch Office
P.O. Box 6770-A
Chicago 80, Illinois
28 August 1945

Dear Dr. Szilard:

Since writing to you yesterday my attention has been called to the contract of employment between the Metallurgical Laboratory, The University of Chicago, and yourself. It is my understanding that you signed this contract on 28 June 1944 and that since that time it has been twice extended, the first extension covering the period 1 July 1944 to 30 June 1945, and the second extension covering the period 1 July 1945 to 30 June 1946.

Your attention is directed to Paragraph 7 of this contract which reads as follows:

"7. It is further understood that you will be bound by and observe all laws, rules and regulations of the United States Government applicable to contracts with respect to the work being carried on and to the disclosure of information with respect thereto. It is also understood that the Laboratory reserves the right and privilege to terminate this contract of employment immediately, for good and sufficient cause, including but not limited to incompetency, neglect of duty, violation of the applicable rules and regulations of the Laboratory or of the United States Government, or conduct inimical to the interests of the United States Government."

In my opinion the portion of the contract quoted above not only outlines certain contractual relations between you and the University, but specifically puts you on knowledge of the necessity for compliance with the regulations of the United States Government concerning disclosures of classified military information. May I repeat what I wrote you in my letter of 27 August, that such commitments as the above preclude the disclosure by you of any information considered secret by the head of the Manhattan District, Major General L. R. Groves.

Sincerely,
James S. Murray
Captain, Corps of Engineers
Intelligence Officer

cc:
LT. Col. W. B. Parsons
Major Claude C. Pierce, Jr.
Capt. J. H. McKinley

Document 116

August 29, 1945

[stamped:]
Metallurgical Laboratory
P.O. Box 5207, Chicago 80, Ill.
Office of the Director
Aug. 30 1945

Dr. R. M. Hutchins
Chancellor
University of Chicago

Dear Dr. Hutchins:

In July of this year, a petition was sent to the President which was signed by 67 scientists employed by the University of Chicago. After the use of the atomic bomb, I advised the Manhattan District that the petition would no longer be treated as "Secret" and obtained the approval of the Manhattan District for this decision. I am now asked by the Manhattan District to reclassify the text of this petition as "Secret."

In a letter dated August 28, 1945, of which I enclose a copy, the Manhattan District asserts that I would be violating my employment agreement with the University of Chicago if I disclosed the text of the petition.

The Manhattan District's definition of "Secret" includes "information that might be injurious to the prestige of any governmental activity," which is, of course, very different from the definition adopted by Congress in passing the Espionage Act.

A separate telephone call received last Saturday from the Manhattan District advised me that I might be violating my employment contract with the University if I were to publish any article or release anything to the press relating to the problems arising out of the development of the atomic bomb, without first obtaining the approval of the Manhattan District.

The unauthorized disclosure of any scientific or technical information which is in fact secret is, of course, covered by the Espionage Act.

It so happens that I personally have been persistently opposed even to the authorized release of such information at the present time and warned against the publication of the Smyth Report[40] as inconsistent with the attitude taken on other related issues by the government.

It so happens that I personally believe that we should all fully cooperate with the Government at present since it is presumably striving to negotiate some international arrangement aimed at the control of the manufacture of atomic bombs. This cooperation on our part might require restraint in the public utterances of the scientists who have been connected with this development. Such restraint, however, must be voluntary and cannot be successfully achieved by coercion.

Coercion in this respect ought, in my opinion, to be resisted by the scientists and I, for one, am not willing to submit to it.

There was no intention of releasing for publication the text of the petition without checking first with the White House, and I have in fact been in communication with the White House on this subject. Apart from a possible release of the text of the petition, I have not contemplated sending any articles to magazines or making any releases to the press.

When I signed my employment contract with the University I was not aware of the possibility that this contract might be interpreted along the lines now indicated to me by the Manhattan District. The purpose of this letter is to raise the question whether the University intends to take the position that my contract implies the restriction of my freedom of action which the Manhattan District thinks it does.

Does the University intend to take the position that I would violate the contract with the University if I made the text of the petition public, assuming that the text contains no disclosure of secret scientific or technical information or anything which, according to the definition of our laws, is in fact a military secret, and taking into consideration that the Manhattan District now chooses to consider the text of this petition as "Secret"?

If I wrote articles or made press releases without previous approval of the Man-

---

[40]Henry D. Smyth, *Atomic Energy for Military Purposes. The official report on the development of the atomic bomb under the auspices of the United States government, 1940–1945* (Princeton: Princeton Univ. Press, 1945).

hattan District, would the University wish to take the position that my action would violate my employment contract, assuming that those articles did not contain any technical, scientific or military information which is in fact secret but were considered, on other grounds, undesirable or "secret" by the Manhattan District?

I regret to have to raise this issue and take your time and attention, but I hope that you will consider this special case as part of the broader issue which is involved.

Very sincerely yours,
Leo Szilard

Chapter VII
It was possible to tell people what we are facing in this century.

Immediately after Hiroshima, I went to see Hutchins and told him that something needed to be done to get thoughtful and influential people to think about what the bomb might mean to the world, and how the world and America could adjust to its existence. I proposed that the University of Chicago call a three-day meeting and assemble about twenty-five of the best men to discuss the subject. Hutchins immediately acted on this proposal, and he invited a broad spectrum of Americans ranging from Henry Wallace to Charles Lindbergh. Lilienthal attended this meeting; so did Chester Barnard, Beardsley Ruml, and Jake Viner.[1] (*Docs. 119–120*)

This was one of the best meetings that I ever attended. In a short period of time we discussed a variety of subjects. We discussed the possibility of preventive war; we discussed the possibility of setting up international control of atomic energy, involving inspection. The wisest remarks that were made at this meeting were made by Jake Viner, and what he said was this: "None of these things will happen. There will be no preventive war, and there will be no international agreement involving inspection. America will be in [sole] possession for a number of years, and the bomb will exert a certain subtle influence; it will be present at every diplomatic conference in the consciousness of the participants and will exert its effect. Then, sooner or later, Russia also will have the bomb, and then a new equilibrium will establish itself." He had certainly more foresight than the rest of us, though it is not clear whether what we have now is an equilibrium or whether it is something else.

One of those who attended the Chicago meeting was Edward Condon. Henry Wallace was at that time looking around for a director for the Bureau of Standards, because Lyman J. Briggs had reached the retirement age. I asked that Condon be invited, with the possibility in mind that he might be a suitable candidate. Wallace liked him at first sight, and Condon was interested in the position. What I did not know when I thought of Condon as a suitable candidate was the fact that Condon had admired Henry Wallace for a number of years. After the conference I had a discussion with Hutchins and Condon, and I proposed that Condon and I go to Washington for a few days and try to find out what the thinking in Washington about the bomb might be.

William Benton, vice president of the University of Chicago, had just accepted an appointment as Assistant Secretary of State under Byrnes. When he heard that we were going down to Washington he offered to invite the top desk men of the State

---

[1] Henry Wallace, Secretary of Commerce; David E. Lilienthal, head of the Tennessee Valley Authority; Chester I. Barnard, Bell Telephone Company executive and government consultant; Beardsley Ruml, treasurer of R. H. Macy and Son and chairman of the Federal Reserve Bank of New York; Jacob Viner, University of Chicago economist. The conference, held September 19–22, involved some fifty participants. Lilienthal published his notes of the meetings, including Szilard's remarks (September 21) as Appendix A, *The Journal of David E. Lilienthal, Vol. II: The Atomic Energy Years, 1945–1950* (New York: Harper & Row, 1964), pp. 637–645.

Department to dinner, and he asked whether Condon and I might give a short discourse on the bomb for the benefit of the Department of State. This we actually did, and I think that this was the first intimation that these people in Washington had, that the advent of the atomic bomb did not necessarily mean that American military power would be enhanced for an indefinite period of time.

I realized very quickly in Washington that for the time being at least the scientists who were regarded as being responsible for the creation of the bomb had the ear of the statesmen. It seemed reasonable to believe that the Russian government, which at that time was more than ever dependent on the cooperation of the scientists for the development of atomic energy, would also be willing to listen to whatever Russian scientists might have to say to them. Therefore I proposed to Mr. Benton that we try to arrange for a conference between Russian and American scientists and that we try to reach a meeting of the minds on what must be done to safeguard the world. . . . I thought that the Russian scientists and we would understand each other and that we had a much better chance to reach an agreement with Russia if discussions with American and Russian scientists were to precede any negotiations between the two governments. (*Doc. 122*) Mr. Benton was enthusiastic about this proposal but Byrnes, who was at that time Secretary of State, vetoed the proposal. In the meantime I had discussed the general idea with a number of my colleagues who were not willing to admit defeat so easily and were determined to appeal to the President over this issue. Mr. Hutchins, Chancellor of the University of Chicago, saw the President in the presence of the Secretary of State, but again Byrnes vetoed the idea and President Truman did not overrule the Secretary of State. [*14*]

While we were in Washington, we somehow picked up a copy of a proposed bill on the control of atomic energy which the War Department had prepared, and which went under the name of the May-Johnson bill. I took this bill back home with me to Chicago and gave it to Edward Levi of the Chicago Law School to read, who promptly informed me that this was a terrible bill and we had better do something to stop its passage.

While I was in Chicago I read in the newspaper that the House Military Affairs Committee had held a hearing on the bill which lasted for a day, and then they closed the hearing and prepared to report out the bill [October 9, 1945]. At that one-day hearing the proponents of the bill testified for the bill, but no opponent of the bill was heard.[2] This was disquieting news, but I doubt very much that I would have swung into action had it not been for a more or less accidental circumstance.

When the war ended, we were asked not to discuss publicly the bomb. We were under the impression that this request was made because there were some important international negotiations on the control of atomic energy under way, and any public

[2]Secretary of War Patterson, General Groves, Vannevar Bush, and James B. Conant testified. U. S. Congress, House of Representatives, Military Affairs Committee, *Atomic Energy, Hearings on H. R. 4280*, 79th Cong., 1st sess., 1–70 (October 9 and 18, 1945), hereinafter referred to as *House Atomic Energy Hearings*.

discussion at this point could have disturbed these negotiations. We were not actually told this, but we were permitted to infer this, and having inferred it we all decided to comply. Therefore all of us refused the numerous requests to speak over the radio or before groups on what the atomic bomb was and what it might mean to the world. We kept silent. S. K. Allison was the only one who gave a speech, and he said that he hoped very much that the secrecy which had been imposed upon this type of work during the war would be lifted after the war; otherwise, he said, he personally would cease to work on atomic energy and he would start to work on the color of butterflies.[3]

When his speech became known, Colonel Nichols flew from Oak Ridge to Chicago and gathered a number of physicists and asked them just for a little while to be quiet and not to stir up things. There was a bill being prepared, he said, on the control of atomic energy, and when that bill was introduced in Congress that would be the right time to discuss these matters. Hearings would be held, and everyone would have an opportunity to appear as a witness and to have his say. (*Doc. 119*)

On the day when the one-day hearing was held before the House Military Affairs Committee and the hearings were closed, A. H. Compton arrived in Chicago and he met with the members of the project. He told us on that occasion that the War Department had prepared a bill for passage through Congress, and that the request which was addressed to us to refrain from publicly speaking on the subject of the atomic bomb was due to the War Department's desire to pass this law without unnecessary discussions in Congress. I remember that I got mad at this point, and got up and said that no bill on the control of atomic energy would be passed in Congress without discussion if I could possibly help it.

Through pure chance I received a telephone call the next morning from Hutchins, who had lunched the previous day with Marshall Field, and he asked whether I would be willing to talk to somebody from the Chicago *Sun*. I said that I was eager to talk to the *Sun*, but I would not want to talk to the *Sun* without also talking to the Chicago *Tribune*, and would Hutchins call up Colonel McCormick and have somebody from the Chicago *Tribune* come and see me?[4]

In two separate interviews I told the reporters who came to see me that there was an attempt on the part of the Army to pass a bill through Congress without "unnecessary discussions," and the physicists would see to it that this would not happen. Because the information came from Compton and I regarded it as confidential, I did not feel free to identify either myself or Compton in this context, and the Chicago *Tribune* told me that under these circumstances they could not use the story. The Chicago *Sun*, being a less well-run newspaper, did not care, and printed the story on its front page.[5] In retrospect, I know that I made a mistake, and I should have permitted the papers to

---

[3]Samuel K. Allison, physicist from Los Alamos and newly appointed director of the Institute for Nuclear Studies. He gave "Sam's butterfly speech" September 1, 1945. Alice Smith, *A Peril and A Hope*, p. 88.
[4]Marshall Field, publisher of the *Sun*; Robert R. McCormick, publisher and editor of the *Tribune*.
[5]Chicago *Sun*, October 11, 1945.

use my identity and have the story printed both in the *Tribune* and the Chicago *Sun*. But in any case, the fight was on.

I went back to Hutchins and called up Condon, who at that time was associate director of research of Westinghouse, and Condon and I once more went down to Washington to see what we could do. We could probably have done very little had it not been for the excellent advice which we received from Bob Lamb, who was at that time legislative advisor of the CIO.[6] He was recommended to us very highly by a number of people, and even though we did not like the idea of working with somebody who was legislative advisor of the CIO, because we did not want to involve the CIO, we decided to overlook this for the sake of getting really first-class advice.

I don't think that anyone knew the Congress as well at that time as did Bob Lamb. When he read the bill, he agreed with us that this bill must not pass. He arranged for us to see Chet Holifield and George Outland. Chet Holifield was on the House Military Affairs Committee and was picked by Bob Lamb for that reason; George Outland was a friend of Chet Holifield and a highly intelligent and competent Congressman. Both Condon and I went to see these two gentlemen and explained to them what the situation was. In the evening Bob Lamb reported to us that they were convinced that we had a good case, and that Chet Holifield would fight for us. Chet Holifield then arranged for Condon and me to see the chairman of the House Military Affairs Committee, [Andrew J.] May, and Sparkman.[7] He himself joined us at this conversation, and we presented the case to them. May was not impressed, and he shortly thereafter made it public that he was not going to reopen the hearing even though Dr. Condon and Dr. Szilard had asked him to do so.

By this time, however, the scientists in the project got organized in Chicago, in Oak Ridge, and in Los Alamos. Both Chicago and Oak Ridge came to the conclusion that the May-Johnson bill was a bad bill which must not pass, and they were so vocal about it that a larger and larger portion of the press got interested in the fight. Los Alamos, under the influence of Oppenheimer, took the opposite position, and was in favor of the passage of the bill. [2] Oppenheimer was told, and he believed it, that no international control of atomic energy could be negotiated until Congress had passed a bill for the control of atomic energy. Therefore, thinking that international control of atomic energy was more important than anything else, Oppenheimer was all in favor of passing a bill fast—any bill as long as it was a bill. For a while Los Alamos followed Oppenheimer. [10]

Condon and I found that everybody in Washington was greatly interested in the issue. We set ourselves a schedule: everybody wanted to see us, and we decided that we would keep Cabinet members waiting for one day, Senators for two days, and Congressmen for three days before we'd give them an appointment. [2]

As Dr. Condon and I found as we went around Washington, no one was really very

---

[6]Robert K. Lamb, Congress of Industrial Organizations.
[7]John J. Sparkman, a member of the committee.

much in favor of the bill except the War Department. The Navy didn't care very much for it and the Interior didn't think much of it. The Department of Commerce thought the bill was bad. The White House had not taken a position even though for a while the impression was created that the bill was an Administration bill. The Office of War Mobilization and Reconversion, which was in charge of steering bills through Congress, didn't like the bill at all. We did not have very much more to do than to tell everybody what everybody else thought of the bill in order to kill the bill. [*10*]

Henry Wallace was very much interested, and he arranged for us to meet Senator Lister Hill.

We went to see Ickes[8] and Ickes grumbled that he had not read this bill at all. The War Department brought it over, left it there for half a day, and then took it away again. "This is not the first time," he said, "that Royall[9] has given me the bum's rush."

We went to see Lewis Strauss who was at that time in the Department of the Navy, and discovered that the Navy did not have any particular views about this bill. The bill was prepared in the War Department, and even though the President made some friendly remarks about the bill it was not really in any sense an Administration bill. It was a War Department bill.

We then went to see James Newman, in Snyder's office,[10] which was supposed to steer the bill through Congress. James Newman had read the bill, and when we saw him he said to us, "'I don't believe that you really understand this bill.' "Well," we said, "we didn't really claim to understand it, but we just didn't think it was a good bill." "Well, I don't think it's a good bill either," said Newman, "but I doubt that you understand what it says. Look," he said, "here the bill says: 'There will be a managing director and an assistant managing director, and the managing director has to keep the assistant managing director informed at all times.' Now," said Newman, "have you ever seen a provision of this type in a bill? What does this mean? Clearly, it means that the managing director will be someone from the Army and the assistant managing director will be someone from the Navy, and since the Navy and the Army don't talk to each other, you have to write into the bill that they must talk to each other on this occasion." For all I know it might well be that he was right.

Under public pressure May, the chairman of the Military Affairs Committee, was in the end forced to reopen the hearings. He reopened the hearings just for one more day. Toward six in the evening I received a telephone call from the office of the Military Affairs Committee, asking me whether I could testify before the committee the next morning. I said that I would testify. Who else could testify? There was no one in town whom I knew who had anything to do with atomic energy except Herbert Anderson,

---

[8]Harold L. Ickes, Secretary of the Interior.
[9]Brigadier General Kenneth C. Royall, coauthor with William L. Marbury of the May-Johnson bill.
[10]James R. Newman, a lawyer with a broad background in science, assistant to John W. Snyder, director of the Office of War Mobilization and Reconversion. Newman acted as a science advisor to Truman when the President gave OWMR responsibility for atomic energy legislation. See Alice Smith, *A Peril and A Hope*, pp. 137–138.

who had worked on the project mainly as Fermi's assistant. He was a spirited young man at the time. I asked Anderson whether he was willing to testify and he said he would, so I gave his name to the committee. The War Department asked Oppenheimer and A. H. Compton to testify for the bill, and so there were four witnesses.

I worked through the night and ended up with some sort of a prepared testimony, which I delivered, and I was then questioned by members of the committee. Herbert Anderson testified after me and then came Compton and Oppenheimer.[11] Neither Compton nor Oppenheimer were really, at heart, in favor of the bill. Oppenheimer managed to give the most brilliant performance on this occasion, for he gave the members of the committee the impression that he was in favor of the bill, and the audience, mostly composed of physicists, his colleagues, the impression that he was against the bill. He did that by the simple expedient of answering a question put to him by a member of the committee. He was asked, "Dr. Oppenheimer, are you in favor of this bill?" And he answered, "Dr. Bush is in favor of this bill, and Dr. Conant is in favor of the bill, and I have a very high regard for both of these gentlemen." To the members of the committee this meant that he favored the bill; to the audience composed of physicists this meant that he did not favor the bill.

H. C. Urey was ready to testify, and this was communicated to the chairman, but he was not called. After my testimony, the chairman dryly remarked that I had consumed two and one half hours of the committee's time.[12] It was obvious that the chairman played ball with the War Department and the committee was stacked against us. There was no hope of inducing the committee into amending the bill; but even if there had been some hope, it was not possible to get a good bill by writing a bad bill and amending it. The only hope was to have the bill bottled up in the Rules Committee, and in this we succeeded. The bill never reached the floor of the House.

One of the men whom I saw rather late in the game was Judge Samuel Rosenman in the White House. There was no need to convince Rosenman. "I told the President," Judge Rosenman told me, "that it looks as if the Army wants to pass this bill by number only."

The Senate set up a Committee on Atomic Energy under the chairmanship of [Brien] McMahon, and this committee started hearings on atomic energy legislation early in 1946. They heard a number of witnesses, and when I testified before this committee, delivering a carefully prepared testimony, I found a much friendlier reception than I had found before the House Military Affairs Committee.[13]

In retrospect it seems to me that at this point I could have left Washington, because there was not very much more that I needed to do. There were plenty of other people

---

[11]For others' testimony and Urey's statement see *House Atomic Energy Hearings*, pp. 96–106.
[12]May actually said: "Thank you very much sir. You have consumed 1 hour and 40 minutes, and we are very glad to have you testify, and I think we have been very much informed by your testimony." *House Atomic Energy Hearings*, p. 96.
[13]The hearings ran from November 27 through February 15; Szilard testified December 10. U.S. Congress, Senate, *Atomic Energy, Hearings before the Special Committee on Atomic Energy*, 79th Cong., 1st sess., pursuant to S. Res. 179 (1945–1946); hereinafter *Senate Atomic Energy Hearings*.

interested who were more influential than I was, yet I stayed throughout most of the hearings and listened to the testimony of several distinguished witnesses. One of the most impressive of these testimonies was that of Langmuir.[14]

One of the things which we tried to get across, and tried to get across very hard, was the notion that it would not take Russia more than five years to develop an atomic bomb also. Even though all younger men and everybody who had a creative part in the development of atomic energy were of that opinion, this was a case of "youth did not prevail."

In his book *Speaking Frankly* James Byrnes relates that when he became secretary of state he tried to find out how long it would take Russia to develop a bomb.[15] He needed this information in order to evaluate proposals for the control of atomic energy. He reports in his book that from the best information which he could gather he concluded that it would take Russia seven to fifteen years to make the bomb. He adds in his book that this estimate was based on the assumption that postwar recovery would be faster than it actually was, and therefore, he said, he thought that this estimate ought to be revised upward rather than downward. Dr. Conant, Dr. Bush, and Dr. Compton all estimated that it would take Russia perhaps fifteen years to make the bomb. Why this should be so is not clear, though it is of course possible to contrive a psychological explanation for these overestimates.

The testimony which Dr. Bush gave before the McMahon Committee will always be engraved in my memory. He was asked whether it would be possible to build ballistic missiles which would fly across the Atlantic, or a distance of 3,500 or 4,000 miles; and he said that this was utterly impossible. One of the Senators then turned to him and said, "Now, Dr. Bush, would you say that 3,000 miles is impossible?" Bush said 3,000 miles was conceivable but 4,000 miles was utterly impossible.[16]

I know, of course, very well why Dr. Bush took this position. If you are an expert, you believe that you are in possession of the truth, and since you know so much, you are unwilling to make allowances for unforeseen developments. This is, I think, what happened in this case. [2]

I have been asked whether I would agree that the tragedy of the scientist is that he is able to bring about great advances in our knowledge, which mankind may then proceed to use for purposes of destruction. My answer is that this is not the tragedy of the scientist; it is the tragedy of mankind. [15]

---

[14]Irving Langmuir, physicist at the General Electric laboratories. *Senate Atomic Energy Hearings*, pp. 109–143.
[15](New York: Harper, 1947). On p. 261 Byrnes says he was told it would take seven to ten years.
[16]In fact Bush volunteered that a 2,500-mile range rocket was conceivable; "But 3,000 miles? . . . I think we can leave that out of our thinking. I wish the American public would leave that out of their thinking." *Senate Atomic Energy Hearings*, pp. 179–180.

Documents from August 1945 through December 1945

Szilard's reaction to the destruction of Hiroshima and Nagasaki was intense, as attested by the following letter to a minister and draft of a petition to President Truman:

Document 117

August 11, 1945

Mr. Alfred W. Painter
Rockefeller Memorial Chapel
University of Chicago
Chicago, Illinois

Dear Mr. Painter:

Yesterday, we agreed that I should write you to put on record something about the matters which we discussed.

Presumably, if the war should end within the next few days, there would be a service in your chapel for the students of the University of Chicago similar to the service which was held after V-E day when Mr. Hutchins spoke. I wondered whether you thought that provisions could be made in this service for a special prayer to be said for the dead of Hiroshima and Nagasaki. If such a prayer were scheduled, this fact could perhaps be stated on the handbills announcing the service which may be distributed to the students so that those who would object could stay away.

I also wondered whether it would be possible to arrange for an offering at the end of the service for the survivors of Hiroshima and Nagasaki with the idea of transmitting the collected sum to the survivors when conditions permitted such transmittal. If this is too difficult to arrange, it would perhaps be possible to suggest to those who attend the service that donations for this purpose be sent to the Swiss Legation in Washington, D.C., for transmittal.

Knowing more about atomic bombs than about church matters, I wonder if any of these suggestions appear desirable and feasible to you.

I understand from what you told me that Mr. Hutchins and Mr. Kimpton would be those primarily concerned with the decision of holding a service at the termination of the war and I am, therefore, transmitting copies of this letter to them.

Very sincerely yours,
Leo Szilard

LS: sw
cc:
R. M. Hutchins
L. A. Kimpton

Document 118

August 13, 1945[1]

To:
The President of the United States
The White House
Washington, D. C.

We, the undersigned scientists engaged in war research at Chicago, believe that further bombings of the civilian population of Japan would be a flagrant violation of our own moral standards. Our nation went clearly and unmistakably on record against this kind of warfare at the time when the Germans bombed the cities of England, and their action was universally condemned by American public opinion. For some time now, our Air Forces have waged a similar type of warfare against the cities of Japan, but not until the use of atomic bombs did the people of this nation fully realize this to be the present policy of our Air Forces. If now, after Japan has expressed in principle her willingness to surrender, we continued to wage war on her civilian population, we would irreparably damage our moral position in our own eyes if not in the eyes of the world.

While Szilard was waiting for the Congressional hearings on control of atomic energy, promised by General Groves' assistant Colonel Nichols, he wrote the following document in preparation. He sent it to John Simpson, H. H. Goldsmith and David Hill in Chicago, and Charles Coryell at Oak Ridge. Commenting on this paper Alice K. Smith writes, "Much of the contents would reappear both in the platform of the atomic scientists and in atomic energy policy as finally adopted. . . . It is virtually the only evidence that the atomic scientists were thinking concretely about domestic control at this time. . ." before they learned the terms of the May-Johnson bill.[2]

Document 119

September 7, 1945

An Attempt to Define the Platform for our Conversations with Members of the Senate and House of Representatives

It should be kept in mind that all contacts with Congressmen and Senators will be arranged primarily for the purpose of discussing bills introduced in Congress. Such contacts will, however, naturally lead to the discussion of the effect of atomic bombs on our relations with other nations.

---

[1] The Japanese accepted American surrender terms August 14, 1945; this petition was never sent.
[2] *A Peril and A Hope*, pp. 91–92. Hewlett and Anderson, in *The New World*, p. 422, incorrectly identify John Simpson as the author of this paper.

We do not at present know what kind of bills will be seriously discussed in Congress. However, it is possible to make certain guesses concerning the wishes of the Interim Committee in which Dr. Bush, J. B. Conant and Dr. A. H. Compton were suggested as experts under the chairmanship of the Secretary of War.

It is quite possible that bills will be introduced in an attempt to keep the secret of the atomic bomb by introducing some sort of a new espionage act. I believe that we should take the following stand:

1. Any set-up which is proposed for the control of atomic power within the United States ought to be opposed if it is likely to create vested interests which will make it more difficult to reach agreements with other nations on the basis of reciprocity.

2. While scientists do not like secrecy, we realize that we might glide, much against our wishes, into an armament race. If such were the situation, we would be willing to work under conditions under which knowledge essential for the further progress of this work would be kept secret.

This has been done and could be done again by the voluntary cooperation of the scientists and any attempt to try to stop leaks by passing some kind of reinforced espionage act would have undesirable consequences.

In the past we had to deal with two aspects of secrecy. We had to keep our results secret from the enemy and we were ordered to keep our results secret from each other. This latter kind of secrecy was called compartmentalization of information. Had we observed the letter of the regulations relating to compartmentalization of information, we would have sabotaged our work and there were a great number of patriotic violations of these rules of secrecy, i.e., unauthorized disclosure of information in the best interest of our work.

If a reinforced espionage act were now to be passed, the results would be that a number of us would simply quit working for the Government. The rest of us who would be willing to continue under the new law would have to choose between obeying the rules and thereby slowing down the work, or violating the rules and thereby offending the law.

I think most of us would, in such circumstances, choose to violate the law. Such violations of the law would, of course, not be prosecuted by the Atomic Power Commission. But it would create an intolerable situation in which the scientists involved could be intimidated and could not openly raise their voices in criticism of the management of the Atomic Power Commission without incurring the risk of being prosecuted for violation of the new espionage act.

3. If an Atomic Power Commission is created, I believe we should insist that any man who is put in a position of influencing the decisions of the Commission be connected with the work on atomic bombs on a full-time basis. Unpaid advisors whose attentions are largely occupied with subjects not connected with the field of atomic power are an evil and must be considered unacceptable.

4. We might propose the creation of a permanent Congressional Committee to supervise the management of the Atomic Power Commission and the scientists ought to be free, irrespective of any law or administrative order issued by the Atomic Power

Commission, to communicate to members of that Committee such information as they consider relevant.

(This proposal is included in the present draft of the tentative platform but might be stricken from the final draft.)

5. We ought to explain to Congressmen that the biggest secret, i.e., that it is possible to make atomic bombs, has been given away when we detonated our first bomb over Japan. The next important secret constitutes knowing along what ways one might move in order to reach this goal. This secret has been given away by the War Department in releasing the Smyth Report.[3]

There are secrets which so far have not been disclosed and which may be considered important, but they affect the development of atomic bombs beyond the stage of development already achieved rather than the present stage of development.

The Smyth Report communicates to other nations so much knowledge that they are as far as knowledge is concerned no worse off and perhaps somewhat better off than we were in the Fall of 1942. As far as knowledge is concerned, they could easily have atomic bombs available within three years. If a nation were determined to manufacture atomic bombs and does not have atomic bombs in manufacture within three years this would be due not to lack of knowledge but to other causes such as lack of raw materials or lack of scientifically trained personnel.

6. We should make it clear that apart from the danger of war which exists among nations in the absence of both an international police and universally-accepted principles upon which a settlement of disputes could be based, the existence of atomic bombs creates an additional danger of war, i.e., the danger of a preventive war arising from a race in the production of atomic bombs. We should point out that this additional danger of war cannot be removed unless nations are willing to accept a close supervision of many of their activities conducted on their own territory by agents of other nations.

Szilard's first address to a larger public, at the closed conference he had urged the president of the University of Chicago to call, extended the ideas he had developed during the war and expressed them forcefully:

Document 120

Address to Atomic Energy Control Conference,
University of Chicago, September 21, 1945[4]

Part I

Any attempt to formulate what the policies of this nation should be—in the present

---

[3]Smyth, *Atomic Energy for Military Purposes*.
[4]The document is dated September 23, 1945. There is a heading: "Below is, written down from memory, the address given in the morning on Friday, September 21, 1945, at the Atomic Energy Conference at the University of Chicago.
"The purpose of this address was to define the task confronting the subcommittee headed by me which was scheduled to meet in the afternoon."

contingency—must be based on some set of assumptions. I should like to attempt to formulate those of my beliefs which are relevant in this connection. . . .[5]

If we focus our attention on the next twenty-five years we may say that development is likely to reach some point intermediate between the first bomb detonated over Hiroshima and processes which once initiated might put an end to all life on earth. Just what intermediate point will be reached within twenty-five years no one can tell.

Part II

When the Japanese surrendered we did not stop our production of atomic bombs. On what cities do we intend to drop these bombs? Our enemies are defeated but some of our Allies may prove difficult to live with, so we are making atomic bombs and we reserve to ourselves the possibility of using them if worst comes to worst. Perhaps we are resolved not to use them but think that the mere possession of them will impress other countries and may make it easier to settle the difficult international questions which now after the war confront us. This does not seem to me to be sound psychology. In peacetime these bombs will be of no earthly use to us unless we are actually determined to go to war if necessary about some question which we consider vital. If we are not actually willing to go to war we can use this bomb for purposes of bluffing only, and the others will see through our bluff.

Three to six years from now Russia is likely to have a stockpile of atomic bombs of her own. I do not believe that an armed peace in which both the United States and Russia have large stockpiles of atomic bombs can be a durable peace.

Part III

I do not believe that permanent peace can be had at any lesser cost than the cost of a World Government. World Government we cannot have at once—not by peaceful means—because we cannot bring about the necessary shift in the loyalty of people simply by passing laws, national or international.

The only thing that we can hope to establish in the immediate future is therefore a durable peace, but the question is, can we make it durable enough to allow time to reach the ultimate goal of a World Government before another world war breaks upon us. If the peace which we establish is not durable enough, then all we shall have achieved is to postpone the third world war and the later it comes the more complete will be the destruction which it will cause.

If we could insure a period of peace for twenty to thirty years this might give us enough time—provided we are determined to make use of it—to approach step by step, perhaps in accordance with some predetermined fixed schedule, the ultimate goal of a world government.

---

[5]We omit one page in which Szilard warns that much larger atomic bombs can be constructed and speculates on the "ultimate danger" of setting off a nuclear chain reaction which could spread through and consume air, ocean water, or solid land; he said there was now enough knowledge to look into this "remote possibility".

Education alone can hardly be counted upon to bring about the shift of loyalty which is a necessary condition for a stable world government. It will be necessary to create institutions that will actually affect the lives and careers of at least the more highly educated strata of society.

The question is not whether or not we shall have a World Government. World Government is almost a certainty within the period of the next fifty years. If the Germans had won the war we would have a world government right now. If Russia should win the next world war she will certainly extend her government over the surface of the whole earth. If we should win the next war and if we should lose in that war the lives of the 30 million people who live in cities of over 250,000 in this country, perhaps we too will be prepared to take over the government of the world.

It is a priori probable that we shall have a world government only at the cost of a terrible war and if I have to give a personal appraisal for this probability I would put it somewhere near 90%. If we then put forward proposals for the establishment of a durable peace, durable enough to permit a transition without war into permanent peace guaranteed by a World Government, it follows that our proposals can have only a 10% chance of being successful.

Ten percent is not a very high chance, but it seems to me that we have to base our thinking and acting entirely on this narrow margin of hope.

Viewed in this light, our objective must be considered to be a modest one since what we propose cannot be expected to have but a slim chance of success. This fact should make us indulgent towards proposals made by others than ourselves and should make it easier to approach our task in a spirit of humility.

Assistant Secretary of State William Benton, who had been vice-president of the University of Chicago, proved an enthusiastic contact for Szilard in Washington. Benton was unable to attend the Chicago conference, so at his request Szilard summarized the discussions for him in the following letter. Most of the statements in the letter—for example, the call for relocating a large portion of the population of the United States, and the suggestion of a special transnational status for scientists—follow ideas Szilard had brought to the conference and would continue to push forward for some time.

Document 121

October 5, 1945

Mr. William Benton
Assistant Secretary of State
Department of State
Washington, D.C.

Dear Mr. Benton:

In the following I have summarized for your convenience some of the points which were made during the confidential discussion at the University of Chicago.

1. The aim of our policy might be to create a situation in which no atomic bombs

are available and ready for instant use. If we could reach an arrangement with Russia that would give us assurance in this respect (under conditions in which we could be sure that violations would be detected and would become known to the world as further discussed below) we would be in a much better position than if we were engaged in an armament race.

2. From 1919 to 1933 it was reasonable to think of sanctions by an organization like the League of Nations as a possible means of enforcing of arrangements which had been agreed upon between nations. Today with Russia and the United States in a dominating position it is difficult to think of methods of enforcement. It appears therefore advisable to think of arrangements with Russia as agreements which could be legally abrogated by either party at any time.

3. In these circumstances the following question appears to be pertinent: Let us assume that some such arrangement is made with Russia, that this arrangement is extended to all other nations and let us now envisage the possibility that this arrangement is abrogated, say seven or ten years from now, during which time there were no secret violations of the arrangement. If at the time of the abrogation there have been large-scale atomic power installations in operation on the territory of Russia and other countries, how long would it take to convert these installations into factories for atomic bombs and how long would it be until atomic bombs become available in quantity ready for instant use? The answer to this question can be given only very tentatively on the base of certain guesses. The answer is six months to a year on the assumption that certain specific restrictions had been applied previously to the development of atomic power installations. These restrictions would slow down but would not completely inhibit the development of atomic power installations.

4. The abrogation of such an arrangement seven or ten years from now might thus lead within a year to the accumulation of large quantitites of atomic bombs which would threaten the sudden annihilation of all our major cities. The very large concentration of our population in the cities—30,000,000 people live here in cities of over 250,000—makes this country particularly vulnerable.

An arrangement with Russia which can be abrogated legally or otherwise ought therefore to be supplemented by something like a ten-year plan for the relocation of 30 to 70,000,000 people. It is estimated that this would involve an expense of fifteen billion dollars per year for ten years and it appears likely that after the relocation we will still have cities between 100,000 and 500,000, but they might have to be built in certain shapes. Cities one mile wide and fifty miles long with a built-up area of fifty square miles have been discussed and are to be further considered.

6. Perhaps the most serious danger that faces us is the danger of a war which would arise more or less automatically if, in continuation of the present trend, the United [States] and Russia were to compete in piling up large stockpiles of atomic bombs. A war might break out then which neither country really wants.

The suggested arrangement could serve to avoid this danger. The arrangements would not prevent a war if either Russia or the United States actually wanted to go to war with the other. It may be hoped, however, that if for a number of years the arrangement has worked satisfactorily and if there are no strong international tensions there would be no desire on the part of Russia to abrogate the arrangement, knowing that by doing so she would precipitate a race in bomb production which might lead to war at once or within a few years.

7. If we consider seriously entering into such an arrangement, we ought to know in advance what assurances we shall ultimately require within the frame-work of the arrangement in order to be sure that secret violations would be detected. Clearly, for this purpose the arrangement would have to include provisions for inspection of mining operations and certain key points in industry. It may be doubted, however, that inspection carried out by agents of some international agency or by American and Russian agents could offer sufficient assurance.

It would be highly desirable to create conditions in which the native engineers and scientists would be put in a position to act as guardians of the international arrangement and would report violations if they occurred. For the scientists and engineers to play this role, it would be necessary that the various espionage acts insofar as they relate to scientific or technical information, be revoked so that scientists and engineers can be pledged perhaps in the form of a new Hippocratic oath to report violations.

8, If the arrangement would provide for conditions under which practically every Russian scientist or engineer would at least twice a year find himself on a visit in some country outside of Russia together with his family, he would be in a position to report a violation and would then be freed by the international agency from his obligation to return to Russia. Assuming international collaboration in the field of atomic power and assuming that ten or even one percent of the Russian engineers or scientists would live up to their oath, we could be fairly sure that violations of arrangements would be detected and would become known to the world. Keeping track of the scientists and engineers inside Russia appears to be [a] more effective method of inspection than keeping track of the movements of uranium ores.[6]

Very sincerely yours,
Leo Szilard

The following cable from Benton to James B. Conant, who was acting as an adviser to Secretary of State Byrnes at the Moscow "Big Three" foreign ministers conference, takes up a suggestion by Szilard:

Document 122

[William Benton to James Conant] December 22, 1945

HUTCHINS TELEPHONED TODAY QUERYING THE ADVISABILITY OF INVITING FIVE OR TEN

[6]Benton replied (October 12, 1945) that he had been "pondering" Szilard's letter in preparing a memorandum for Secretary of State Byrnes.

RUSSIAN PHYSICISTS TO THE UNITED STATES, BRINGING THEM IN UNDER PRIVATE AUSPICES TO VISIT HARVARD, UNIVERSITY OF CHICAGO, AND OTHER INSTITUTIONS. HE FELT SUCH JOINT DISCUSSIONS OF ATOMIC PHYSICS BETWEEN SCIENTISTS OF THE TWO COUNTRIES MIGHT PROMOTE A BASIS FOR INTERNATIONAL COOPERATION AND CONTROL. SUCH DISCUSSIONS WOULD OF COURSE BE PRIVATE AND UNOFFICIAL WITHOUT GOVERNMENTAL COMMENTS. I TOLD HIM I WOULD CABLE YOU THIS SUGGESTION FOR POSSIBLE EXPLORATION BY YOU IN MOSCOW.

cc:
Mr. Hutchins
Mr. Szilard

We lack space in this volume to document Szilard's extensive post-1945 work to influence Congress and other public figures, create and sustain organizations, and educate public opinion. The preceding cable, for example, represents only the first in a series of proposals Szilard made over nearly twenty years, trying to bring together scientists from different countries in pursuit of world peace. This may help to remind us of his early dreams of a free association of the intelligent and public-spirited and his subsequent efforts to create what he called "a more livable world."

Source Notes
(numbering by order of appearance in text)

[1] Notes, "Rough Draft, Outline for Book," June 1960.
[2] Interview taped in New York, May 1960.
[3] Slightly abridged from interview taped in Washington, D.C., 1963.
[4] "Biographical Data and List of Publications of Leo Szilard from 1922 to 1945," written by Szilard sometime after 1955.
[5] Abridged from ms, "Book. Apology (in lieu of a foreword)," probably 1951.
[6] Draft for magazine article, dictated in 1960.
[7] Draft, Nov. 7, 1953, for a speech delivered at Brandeis, Nov. 19, 1953.
[8] Memorandum to A. H. Compton, Nov. 12, 1942.
[9] Memorandum about the Einstein Letter, April 18, 1955.
[10] Transcript of magnetic recordings made in May 1956. (We do not have the original recordings and have had to guess at the meaning in some cases.)
[11] Slightly abridged from interview given Mike Wallace, WNTA-TV, Feb. 27, 1961.
[12] "The Story of a Petition," July 28, 1946.
[13] Interview in New York, July 29, 1960; printed (with changes and abridgments) in *U.S. News and World Report*, Aug. 15, 1960, p. 68.
[14] Typescript, "This Version of the Facts," transcript of dictated draft of speech for first Pugwash Conference, 1957.
[15] "Answers to Questions," dictated May 9, 1963, probably a condensation of [3].

Name Index

Adams, Walter, 34–35
Adamson, Keith, 84, 86
Allison, Samuel K., 225
Anderson, Herbert L., 55n 56, 81, 87n, 88, 94, 115, 116n, 148n, 168n, 169n, 171, 185n, 227, 228, 231n
Anderson, Oscar E., Jr., 117
Arneson, R. Gordon, 188n
Aschner 40

Bailey, Charles W., 188n
Barnard, Chester I., 223
Barnes, Sidney W., 60
Barrett, William F., 86
  letter from Szilard to, 106–107
Bartky, Walter, 182, 186, 216
Baruch, Bernard M., 92
Batchelor, H. D.
  letter from Szilard to, 150–151
Beams, Jesse, 112, 129, 133
Benton, William, 223–224
  cable to Conant from, 237–238
  letter from Szilard to, 235–237
Bentwich, Norman, 15n
Bethe, Hans, 18, 149
Beveridge, William, 15, 30–31, 32, 34, 112
Biot, Maurice Anthony, 89
Bitter, Francis, 36
Blackett, P.M.S. 17, 18, 57, 70, 71, 72–73, 74, 77
  letter from Szilard to, 77
Boas, Franz, 33
Bohr, Harald, 33
Bohr, Niels, 33
  letter from Szilard to, 44–45
Bose, S. N., 36
Bowen, Harold G., 116, 129, 134, 139
Brailsford, H. N., 22n
Breit, Gregory, 129, 139
  correspondence with Szilard, 133; 135–136
Briggs, Lyman J., 84–85, 122, 129, 130, 134, 139, 140, 143, 152, 159, 176, 177, 205, 223
  letter from Einstein to, 125–126
  letter from Szilard to, 110, 111 See also Briggs Committee
Briggs Committee, 105, 107–110, 112–113, 123, 164
Brown, Allen, 209n, 213n
Bush, Vannevar, 68, 117, 138, 155, 159, 163, 167, 181, 185, 224n, 228, 229, 232
  correspondence with Szilard, 151–153, 161–163
  memorandum from Szilard to, 164–179
Byrnes, James, 183, 185, 186, 205n, 216, 223, 224, 229

Cattier, Jean E. V., 106

Chadwick, James, 46
Chalmers, T. A., 20
Cherwell, Lord. See Lindemann, F. A.
Cockcroft, John D., 45–46
  letter from Szilard to, 46–48
Collie, C. H., 42
Compton, Arthur Holly, 146, 152, 154, 155–156, 158, 159–160, 161, 167–168, 182, 185, 186, 188, 217, 219, 225, 228, 229, 232
  correspondence with Szilard, 214
  memorandum from Szilard to, 160–161
  transmittal letter to Nichols from, 214–215
Compton, Karl T., 92, 103, 104, 112
Conant, James B., 154, 159, 162, 163, 169, 181, 185, 224n, 228, 229, 232
  cable from Benton to, 237–238
Condon, Edward, 223, 226–227
Connelly, Matthew J., 182, 188
  cable to Szilard from, 216
  letter from Szilard to, 215–216
Consodine, W. A., 219
Coombes, J., 18n
Coryell, Charles, 231
Cottrell, James, 47n
Creutz, Edward C., 158, 167, 168, 212
  letter from Szilard to, 212–213
Curie, Irene, 17

Daniels, Farrington, 215
Dee, P. I., 47, 48
Delbrück, Max, 32–33
Dempster, Arthur J., 44
Dirac, P. A., 57, 70, 71–72
Donnan, Frederick G., 32, 33, 40
Drakenfeld, B. F., 67
Dubridge, Lee A., 139
Duisberg, J., 32–33
Dunn, Gano, 92–93
  letter from Szilard to, 102–103

Eckart, Carl, 11n
Edison, Charles, 56
Einstein, Albert, 8, 9, 11, 12, 22, 83, 84, 98, 102, 104, 115, 119, 122, 123, 130, 137, 139, 152, 183, 208, 216
  correspondence with Roosevelt, 84, 94–96, 104–105, 107–108, 205
  correspondence with Sachs, 120–121
  correspondence with Szilard, 90–94, 96–97, 100–102, 107–109
  letter to Briggs from, 125–126
  letter of introduction for Szilard from, 181

Farkas, 104
Feis, Herbert, 186n
Feld, Bernard T., 11n
Fermi, Enrico, 10n, 46, 53, 55, 56–57, 63–64, 67, 68, 69, 74, 75, 81–82, 86, 87–88, 94,

95, 98, 103, 104, 106, 110, 111, 112, 113, 114, 116, 125, 126, 129, 130, 133, 139, 143, 144, 146, 147, 150, 164, 171, 174, 177, 185
  correspondence with Szilard, 133–135
Field, Marshall, 225
Foote, F. G., 158
Franck, James, 186, 187, 209
Frisch, Otto, 19n, 62, 144

Gary, Tom C., 169n
Goldhaber, Maurice, 21, 44, 53n, 57, 60
  letter to Szilard from, 44
Goldsmith, H. H., 231
Gowing, Margaret, 144n
Greenwalt, Crawford H., 169n
Griffiths, J. H. E., 42
Grodzins, Morton, 186n
Groves, Leslie R., 169, 181, 183, 186, 188, 214, 217, 218, 224n, 231
Gunn, Ross, 56, 82, 134
  letter to Szilard from, 89–90

Haber, Fritz, 8
Hafstad, L. R., 66, 67
Hahn, Otto, 20, 44, 62
Halban, Hans von, Jr., 56n, 57, 68n, 70, 71, 73, 115, 121n
Hamister, V. C.,
  letter from Szilard to, 150
Hanstein, H. B., 55n
Hardy, G. H., 33
Hartog, Philip, 31, 32
Hewlett, Richard G., 117, 148n, 168, 169n, 185n, 231n
Heydenburg, N. P., 67
Hilbert, David, 31
Hill, Archibald Vivian, 16, 33
Hill, David, 231
Hill, Lister, 227
Hill, R. D., 53n
Hirst, Hugo, 38–40
Hitler, 13–14
Holifield, Chet, 226
Hoover, Gilbert C., 84
Horthy, Nicolas, 8
Hutchins, Robert M., 182, 223, 224, 225, 226
  letter from Szilard to, 220–222

Ickes, Harold L., 227
Irving, David, 121n

Jastrow, Ignaz, 15
Joliot, Frédéric, 17, 19, 46, 53, 56, 57, 68, 70, 71, 73, 75, 77, 95, 115, 121n, 135
  cable from Szilard to, 78
  cables to Szilard from, 58, 59
  correspondence with Szilard, 54, 69–70, 74, 78–79, 80, 119

Kaempffert, Waldemar, 64n
Kalckar, F., 44

Kapitza, Peter, 35
Kearney, R. B., 65
Kent, Robert H., 87
Kimpton, L. A., 230
Knebel, Fletcher, 188n
Kowarski, Lew, 56n, 68n, 73, 115, 121n
Kraus, Lili, 16n
Kun, Bela, 8

Lamb, Robert K., 226
Langmuir, Irving, 229
Lapp, Ralph, 212
Laski, Harold J., 31
Lawrence, Ernest O., 11, 72–73, 139, 146, 152, 159, 161, 168, 185
Lawson, Andrew W., Jr., 133–134
Levi, Edward, 224
Lewis, Warren K., 169, 173
Liebowitz, Benjamin, 33, 48, 55
  correspondence with Szilard, 113–114
Lilienthal, David E., 223
Lindbergh, Charles, 92, 96, 100, 223
Lindemann, F. A., 21, 41–42, 49, 50
  cable to Szilard, 63
  correspondence with Szilard, 50–52, 192–196
Loomis, A. L., 75

McCormick, Robert R., 225
McKinley, J. H., 217
McMahon, Brien, 228
McMillan, Edwin, 212
MacPherson, H. G., 150
Madach, Imre, 3n
Mallinckrodt family, 167
Mandl, Otto, 16–17
Marbury, William L., 227n
Marschak, Jacob, 14–15
Marshall, John, Jr., 158
May, A. J., 227, 228
Meitner, Lise, 13, 44, 62
Melchett, Henry Mond, 32
Meyer, R. C., 66n
Mitchell, Dana P., 154
Moore, Thomas V., 83, 157, 159, 168
Morrison, Philip, 212
Murphree, Eger V., 146, 152, 154, 159, 168, 169, 179
Murray, James S.
  letter to Szilard from, 216–220
Murray, Mary, 36
Mussolini, 24n

Namier, Lewis Bernstein, 14
Nathan, Otto, 22n
Nernst, Walter, 8
Nichols, K. D., 188, 225, 231
  transmittal letter from Compton to, 214–215
Norden, Heinz, 22n

Oliphant, Marcus, 46, 47, 146
Oppenheimer, Frank, 162, 212, 213

Oppenheimer, J. Robert, 162, 185, 209n, 213, 226, 228
Outland, George, 226

Painter, Alfred W.
  letter from Szilard to, 230
Parsons, W. B., 219, 220
Patterson, R. P. (Secretary of War), 224n
Pegram, George B., 55, 56, 57, 82, 86, 106, 111, 112, 113, 116, 129, 130, 133, 134, 135, 139
  correspondence with Szilard, 109–110, 139–141
Peierls, Rudolf, 144
Perrin, F., 115, 121n
Pierce, Claude C., Jr., 217, 219–220
Placzek, George, 44, 81
Planck, Max, 8, 31
Polanyi, Michael, 13, 19, 40
  correspondence with Szilard, 43
  letter from Weizmann to, 43

Rabi, Isidor Isaac, 54, 57, 75, 87, 117
Rabinowitch, Eugene, 186n
Richards, 75
Roberts, Richard B., 57, 66, 73, 75–76, 85
Roosevelt, Franklin Delano, 90, 101, 116, 119–120, 125, 182
  correspondence with Einstein, 84, 94–96, 107–108, 205
  correspondence with Sachs, 121–122
  letter from Szilard to, 205–207
  memorandum from Szilard to, 181–182, 196–204
Roosevelt, Eleanor, 181
Rosenman, Samuel, 228
Royall, Kenneth C., 227
Ruml, Beardsley, 223
Russell, Edward John, 32, 33
Rutherford, Ernest, 17, 20, 40
  letter from Szilard to, 45–46

Sachs, Alexander, 84, 85, 90, 100, 101, 107–108, 110, 111, 116, 119–120, 122, 125, 126, 130, 134, 139
  correspondence with Einstein, 120–121
  correspondence with Roosevelt, 121–122
  correspondence with Szilard, 97–98, 112–113, 123–125, 137–139
  and Einstein's letter to Roosevelt, 104–105
Schacht, Hjalmar, 13
Schlesinger, Karl, 14–15
Sengier, Edgar, 106, 131
Sherwin, Martin J., 209n
Simon, Francis, 144
Simpson, Esther, 35
Simpson, John, 231
Slade, R. E., 51
Smith, Alice K., 215n, 227n, 231
Smyth, Henry DeWolf, 117n, 148, 154, 221n
Sparkman, John J., 226
Spedding, Frank H., 168

Stimson, Henry L., 185, 188n
Stolper, Gustav, 84, 90
Strassmann, Fritz, 20n
Strauss, Lewis L., 62–65, 70, 86, 227
  correspondence with Szilard, 62, 63–65, 74–75, 77, 78, 88–89
Strauss, Mrs. L., 63, 89
Szilard, Bela, 66n
Szilard, Gertrud Weiss, 11n
Szilard, Leo, 53, 71, 94, 95, 104, 123, 125, 126, 139
  on Academic Assistance Council, 32–36
  address to atomic energy conference, 233–235
  attitude toward atomic bomb tests and use, 181–188, 231–235
  in Austro-Hungarian Army, 6–8
  and Briggs Committee, 107–110
  career of, xi
  childhood of, 3–5
  courses at University of Berlin, 10
  on dangers of Germany getting atomic bombs, 152, 192–196
  on danger of war in 1933, 35–36
  and development of cyclotron, 11
  on discovery of nuclear chain reaction mechanism, 38–48
  efforts to declassify petition on atomic bomb use, 215–222
  efforts to obtain radium, 63, 65
  efforts to present viewpoints to Roosevelt, 95–105, 119–120, 204–207
  efforts to prevent use of atomic bomb, 185–188
  efforts to see Truman, 182–183
  on England, 35, 36
  and funding of chain reaction experiments, 38–43, 55, 105–107, 109–117, 130–131
  and international scientific cooperation, 32–34, 238
  memorandum on peacetime use of atomic energy, 189–192
  and organization of atomic bomb project, 129–131, 133–141, 151–179
  proposal for "Der Bund," 22–30
  proposed scientific boycott of Japan, 36–38
  reaction to destruction of Hiroshima and Nagasaki, 223–238
  reaction to discovery of nuclear fission, 60–65
  and secrecy issue, 118, 126–141
  struggle against May-Johnson bill, 226–228
  at Technische Hochschule of Berlin, 8–9
  at University of Berlin, 9–11
  work and plans for rescuing refugee scientists, 30–34
  during World War I, 4–6

Tate, John T., 73, 128, 132, 133
  correspondence with Szilard, 119–120
Teller, Edward, 56, 63, 66n, 70, 84, 85, 88, 90,

93, 97, 98, 108, 110, 111, 129, 139, 160–161, 209n, 212, 213n
  letters to Szilard, 66, 208–209
Thomson, George Paget, 18, 177
Truman, Harry S, 182–183, 224
  letter from Szilard to, 208
  scientist's petition to, 187–188, 209–215
Tuck, James L., 49
  letter from Szilard to, 48–50
Turin, Lucia, 31
Turner, Louis A., 117, 133
  correspondence with Szilard, 127–129, 132–133
Tuve, Merle A., 56, 57, 66, 67–68, 85, 110, 112, 129, 139
  letter to Szilard from, 68

Urey, Harold C., 117, 131, 132, 133, 134, 139, 144, 152, 159, 161, 164, 168, 172, 173, 176, 177, 178, 179, 183, 216, 228

Vernon, 174
Viner, Jacob, 223
Von Laue, Max, 8, 9, 11

Wallace, Henry, 223, 227
Wang, P., 66n
Watson, Edwin M., 122
Weinberg, Alvin, 199n
Weisskopf, Victor, 57, 70, 77
  letter to Blackett from, 72–73
Weizsaecker, Carl F. von, 95–96, 120–121, 122
Weizmann, Chaim, 18–19 31, 32
  letter to Polyanyi from, 43
Wells, H. G., 16, 17, 18, 22n, 38, 53
Wheeler, John, 137, 171n
Whitaker, Martin D., 215
  letter from Szilard to, 213
Wigner, Eugene D., 53, 54, 55–57, 63, 65, 70, 72, 82–83, 84, 85 90, 98, 101n, 107–108, 111, 117, 121, 123, 125, 129, 133, 139, 147–148, 157, 166, 167, 168, 170, 171, 173, 174, 177, 209, 213
  correspondence with Szilard, 87–88, 97, 103–104, 136–137
  letter to Dirac from, 72
Williams, Roger, 169n
Willstätter, Richard, 43
Wilson, Robert, 212
Wright, C. S., 18n

Young, Gale, 148n

Zinn, Walter, 55, 56, 113